IN SEARCH OF SPARROWS

To

my long-suffering wife, Margaret, who
accompanied me on most of the searches

In Search of Sparrows

by

J. DENIS SUMMERS-SMITH

Wherever are they?

Illustrated by
Euan Dunn

T & A D POYSER

London

First published 1992 by T & AD Poyser Ltd
Print-on-demand and digital editions published 2010 by T & AD Poyser,
an imprint of A&C Black Publishers Ltd, 36 Soho Square, London W1D 3QY

www.acblack.com

ISBN (print) 978-1-4081-3783-3
ISBN (epub) 978-1-4081-3782-6
ISBN (e-pdf) 978-1-4081-3781-9

A CIP catalogue record for this book is available from the British Library

This is a print-on-demand edition produced from an original copy.

It is produced using paper that is made from wood grown in managed sustainable forests. It is natural, renewable and recyclable. The logging and manufacturing processes conform to the environmental regulations of the country of origin.

Printed and bound in Great Britain

FSC www.fsc.org MIX Paper from responsible sources FSC® C013604

Contents

List of illustrations

Preface and Acknowledgements

It is hoped that this account will appeal not only to those interested in birds, but also to those interested in far-away places – the places themselves and the people who live there rather than with man's artefacts as in most travel books.

Many people helped me in my search. I hope those who are mentioned in the text will accept this as acknowledgement of my gratitude. But there are many others – too many to mention, and some whose names I have forgotten or in some cases not even known. They are also included in my thanks.

A number of friends have read parts or the whole of various drafts of the book and I am most grateful to them for constructive criticisms. I have paid close attention to these and am sure that this has resulted in considerable improvement. They have done their best – the remaining deficiencies are entirely mine. In particular, I should like to thank Mrs M. Betton, J. B. Devenport, Dr P. Evans, L. R. Lewis, T. Poyser, Mrs N. Robinson and last, but not least, my wife.

The photograph on p. 89 was taken by Peter Evans, the others by the author or his wife, Margaret.

Finally, my thanks must go to Euan Dunn, whose cartoons so nicely capture the flavour of my adventures in search of sparrows.

Help!

1: Beginnings

Twa birds sat on a barra
Yin was a spug the ither a sparra.

Anon

I do not remember when I first became interested in natural history. I am sure any early interest was stimulated by long childhood holidays in Donegal with an uncle who was a country parson. I do not wish to cast any aspersions on his memory, but I feel that, like many country parsons of that era, his pastoral duties were not unduly onerous. Suffice to say, he taught me to fish and, though he was more of a botanist than an ornithologist, gave me a sound grounding in the local birds.

In the 1930s my interest in birds had to compete with the many others that a growing boy is subject to. Six years in the army during the war provided little opportunity for birdwatching, apart from spells guarding the east coast of England against a possible German attack, when as Intelligence Officer of my battalion with a sympathetic commanding officer I spent some time surveying such likely spots for invasion as the newly flooded Suffolk marshes that had attracted booming Bitterns and the Northumbrian coast with a good mix of seabirds.

It was not until 1947, when I had settled down in north Hampshire, that I decided the time had come when I wanted to make a serious study. This was a time when field ornithology began to take over from what had previously been largely a museum discipline. The change was pioneered by such towering figures as David Lack and Niko Tinbergen, following in the footsteps of Konrad Lorenz and, very conveniently, working in nearby Oxford.

The 1940s was a time of petrol rationing and difficult travel; this led me to choose an accessible species of bird. What could be more accessible than the House Sparrows that were always around in the garden and nested on my house? Here was something that would involve me in the minimum of travel and lost time. It was a fortunate choice. Not only can one find lots of new and unexpected things by making an in-depth study of any particular animal, but, more specifically, postgraduate zoologists, who were embarking on field studies and had the benefit of generous grants, did not want to waste time on the humble sparrows in their back gardens when they could travel to the Galapagos Islands to study Darwin's finches or to Christmas Island for boobies. I virtually had the field to myself and I found much to interest me despite it being one of the most familiar of all birds. Notwithstanding the

Postgraduate students did not want to waste time on the humble sparrows in their back gardens when they could travel to Christmas Island for boobies.

comment by Alfred Newton in his *Dictionary of Birds* published in 1896 that it 'is far too well known to need any description of its appearance or habits', in fact, very little was known about it in detail and it had rather a poor reputation – barely considered as a bird by most birdwatchers. For example, the eminent French ornithologist, the Count de Buffon, writing earlier in the same century, described the House Sparrow in the following terms. 'It is extremely destructive, its plumage is entirely useless, its flesh indifferent, its notes grating to the ear and its familiarity and petulance disgusting.'

After spending six years watching the House Sparrow in my Hampshire village and a further three in a town in County Durham, one thing about the bird began to fascinate me more and more. This was its success as an animal. And I began to consider how I could investigate the reasons for this. The sparrow certainly owes its success to its association with man, the dominant animal species in the world, and the question thus transposes itself to one of enquiring how this association came about. This now becomes an evolutionary question and to get some insight into it I resolved to widen my studies to involve the House Sparrow's closest relatives. The House Sparrow is one of a group of birds that are widely dispersed over the Old World, and, although now to be found virtually everywhere with the exception of Antarctica through introductions by man starting in the 19th century, are best described by the term 'Old World sparrows' to distinguish them from other groups of birds, such as the American or New World sparrows, which are quite unrelated impostors. When I began, received wisdom was that there were 15 species of sparrows, though by 1970 the pundits had increased this to 19 by splitting up the grey-headed sparrows of Africa into 5 separate species and in 1984, before I had finished, I had split off a rufous sparrow, making a grand total of 20 in all. It did not seem an impossible task to have a look at all of them, though this was obviously going to involve some travel. It is somewhat ironic that, having chosen the House Sparrow as a study species for the particular reason that it would not involve me in travel, this should have led me to a worldwide search. At the latest count this has involved some 50 countries, some chosen deliberately in order to see a particular sparrow, others on business trips where advantage could be taken of the now almost omnipresent distribution of sparrows and the opportunity to make observations on the side.

The period starting in the 1950s provided a great opportunity for travel; cheap, or at least relatively cheap, air fares made distant places accessible for visits during my annual leave and the political climate at that time was generally favourable. Air fares continue to be affordable, though travel seems to have become slightly more difficult with the imposition of visa requirements by many countries. I hope the granting of a visa does not become more difficult for me, but if the authorities care to check up they will find that many

Cock House Sparrow. The country male looks really smart in his spring breeding plumage.

of my visits seem to have been followed by revolutions. I was in eastern Czechoslovakia at the time of the Prague Spring when the Russians moved in. Ethiopia, that I visited in 1972, had its revolution later that year. Afghanistan followed in 1978 after my visit in 1976, and Iran, which I visited at the same time, the following year. There was a gap of three years between my visit to the Seychelles in 1974 and the overthrow of the government there in 1977. The situation in the Punjab has not been the same since I visited Amritsar in 1981. In 1984 I spent a week in Hefei, in Anhui Province, in China; the university there was to be the centre for the critical poster upsurge in 1986. While the statistical association seems high, I am sure the relationship is a casual rather than a causal one. It is difficult to imagine how sparrows could have such a political influence, though their supposed depradations on the rice crop led Chairman Mao to launch a rather ill-considered, nationwide campaign against the Tree Sparrow. In the same vein I wonder, if they ever check up on the paperwork that is filled in at borders, what the immigration authorities will think of the journeys I have made by car criss-crossing from France, Switzerland and Austria into Italy at each successive frontier post that one of my searches necessitated.

Unfortunately, whatever the reason, the relatively complete freedom of travel that followed the 1939–45 War seems to have been a passing phase. Many of the new regimes look less favourably on a casual traveller wandering around in out-of-the-way parts of their countries, complete with telescope and cameras with telephoto lenses.

As a small boy I did not follow the usual ambition of wanting to become a train or tram driver. My answer to the inevitable question of what I wanted to become was 'an explorer'. My travels in search of sparrows have simply realized this sublimated desire, though little of it really comes into the classification of exploration. It has, however, had a most rewarding spin-off – I have been privileged to have been welcomed as a traveller or a guest in many countries, avoiding the usual fate of the package tourist of being insulated by the trappings of the international holiday business. In order to make the most of my annual leave, the various sparrow searches were highly planned in advance. This involved writing to local people for information and help. For some reason this seemed to bring out the greatest sympathy – perhaps the feeling that anyone so misguided or unworldly as to be passionately interested in such a common dull bird as a sparrow must obviously be in need of support and guidance – and awakened a protective instinct. I do not know, but whatever the reason, I received not only useful information but invitations to stay, with the result that I have made friends with many people that otherwise it is most unlikely that I should ever have met. I can recommend nothing better than a specialized interest in an obscure subject for raising people's curiosity and effecting an introduction. After that it's up to you.

As we shall see, many sparrows in addition to the House Sparrow have a close association with man, and this means that much of the watching has to be carried out in built-up surroundings. The best time for birdwatching is in the early morning hours and this has led to problems. People do not appear to like having someone watch their houses through binoculars at that time of day and twice I have been questioned by the police about my activities. Sparrows also occur regularly in farmland near towns and another time I was treated roughly by a farmer who took me for a Peeping Tom. Nowadays, a telescope, permanently mounted on a large tripod, is *de rigueur* as far as the birdwatcher is concerned; when I started watching birds somebody carrying field glasses was more likely to be on their way to a race meeting or even be a Peeping Tom. These minor problems were in England where there were no language difficulties in explaining the situation. On another occasion in the Spanish Pyrenees I was less sanguine. I was skulking along a hedge early in the morning shortly after dawn when I bumped into an army patrol. This was in the Valle d'Aran in the late 1950s when birdwatching was scarcely recognized and I feared the worst. My Spanish at that time was limited to little more than *buenos dias,* but even that seemed enough to do the trick. I found out subsequently they were out looking for smugglers from Andorra and, although I had been there for only a few days, the presence of the mad Englishman was already well known in the valley (despite being a Scot it sometimes pays to hide under the cloak of the mad Englishman abroad, keeping the card of one's true identity up one's sleeve to be played later!). Years later when I was lecturing (on engineering) in Venezuela my interpreter enquired why I watched birds. A quick response seemed to satisfy her: I said I liked looking at pretty things, but if I looked at the girls, particularly through binoculars, their boyfriends or husbands were liable to object physically; nobody seemed to mind if I confined my attention to the feathered birds. Birdwatching was obviously such an unfamiliar activity there that she was unable to give me a Spanish equivalent for the term.

Borders are not respected by birds and present particular problems for the birdwatcher. Once in Austria near the Hungarian border through bad map reading I emerged from the cover of a hedge to find a view in my binoculars of a soldier in a border watchtower with a rifle trained on me. I gave him a cheery wave and beat a hasty retreat. On another occasion I drove by mistake into the zone between Albania and Yugoslavia on the shore of Lake Ohrid. We were immediately surrounded by an armed patrol with bayonets drawn, but things were quickly sorted out once I had agreed to confine my watching to the Yugoslav birds. In some of the Middle East countries birdwatchers have been less lucky and have ended up in jail under suspicion of spying. The only encounter I have had with the authorities in that area has been with the border guards in Israel. Israel is such a small country that one always seems to be bumping into the frontier. I have, however, found the Israeli soldiers to be most courteous, merely like the Yugoslavs requesting that birdwatching and photography be confined to their side!

Although the search has largely been a physical one, it has also been one of the mind. I have searched avidly for anything written about sparrows and when this has been in a language inaccessible to me I have badgered friends and even casual acquaintances to translate it into English for me. Again, I have been surprised and grateful for their tolerance as this has extended not only to western European languages, but to Russian, Hebrew, Mandarin and even Georgian, overcoming an inaccessibility no less formidable than the geographic ones that had to be surmounted.

Sparrows are mentioned several times in the Bible, though whether these always refer to the birds we now recognize as such, or merely to any small bird found around houses, is not clear. The word sparrow apparently comes from the Old English *speerwa* and represents the archetypal little brown job, or LBJ, of the modern birdwatcher. Different types of sparrows were, however, identified in Classical times and Alexander of Myndos apparently distinguished between domestic (House?) and wild (the modern Tree, Willow or Rock?) sparrows in the 1st century AD, and the Latin word *passer* probably referred to the bird we know today as the House Sparrow, though this must remain somewhat speculative. (The Willow Sparrow is also known as the Spanish Sparrow. I prefer the name I have used in the text as the bird is frequently associated with willows and is by no means common in Spain.)

This book is an account of the journeys that I have undertaken; most travel books are obsessed about man's artefacts, or are accounts of difficult journeys undergone specifically because of that difficulty and the enjoyment of the hazards and unpredictability. This one is concerned with the search for a group of birds making up less than 0.2% of the bird species in the world, though in searching for them I have now seen over 20% of the total number. It is intended as a tribute to all those who have so unstintingly given me their help, both those mentioned in the text and those not, to whom, if they chance to read it, it might trigger off a happy (I hope) recollection. They have all done so much, not only to have allowed me to follow my narrow interest in sparrows, but also to have broadened my outlook. I thank them one and all.

Dispute over the title 'house' sparrow.

2: Which is the 'house' sparrow? Afghanistan

Hark, 'tis the sparrow's good-night twitter
About your cottage eaves!
Robert Browning, '*The lost mistress*', 1845

The bird we call the Tree Sparrow gets its name from the fact that its preferred nesting site in Britain is in holes in trees. It is not a woodland bird, but is to be found more where there are scattered trees, particularly hedgerow trees bordering fields; in Germany its name *Feldsperling* (= field sparrow) recognizes this habitat. However, in the Far East (and there is reason to believe that the Tree Sparrow evolved in southeast China in association with sedentary agricultural man there in the same way that I consider the House Sparrow evolved in the Tigris–Euphrates valley), where the House Sparrow does not occur, the Tree Sparrow is a complete 'house' sparrow, living in inhabited places and nesting in holes in houses in exactly the same way that the House Sparrow does with us in Europe. It is thus interesting to see what occurs where these two 'house' sparrows first come into contact.

One such place is Turkestan in Central Asia, partly in the USSR and partly administered by China. This area was traversed in 1907–8 by Captain D. Carruthers of the Indian Army, who on one of his leaves 'walked' home from the North-West Frontier, over the high mountain passes through southeast Asia and on into Europe. His book describing this journey is a fascinating one and shows how much freer travel was in those days with no need for passports and visas, provided one had the necessary spirit of adventure and a bag of gold to defray the expenses. Captain Carruthers was moreover an excellent naturalist and there are many references to the birds and animals he encountered on the way. During a delay in Kara Kul in Turkestan, because of lack of transport, he spent the time paying particular attention to the local sparrows. He found the Tree Sparrow to be the common sparrow of the towns and cultivated areas, whereas the House Sparrow was more of a country bird – quite a reversal of the roles one would expect from having seen the birds in Europe. In fact, the Tree Sparrow is a resident species in this area, while the House Sparrow is only a summer visitor, crossing over the mountains to winter in the plains of northern Pakistan and India. It is evident that the House Sparrow has spread north from the Indian subcontinent into the high plateau of Afghanistan and has been forced to adopt a migratory role as it was unable to withstand the severe climate of this elevated region. The spread must have eventually taken the bird right over the mountains into the Turkestan basin, where the climate would have allowed it to overwinter, but despite this the habit of migration has been maintained. It would seem that the resident Tree Sparrows, which must have been there before the House Sparrows arrived (probably about 2000 years ago), have been able to maintain their status as 'house' sparrows over their larger and heavier congeners (typical weights for the sparrows living in the area are: House Sparrow 25–30 grams, Tree Sparrow 20–25 grams) by already being in possession of the favoured nesting holes in buildings before the arrival of their migratory cousins in the spring. The story does not, however, end there. In the 19th century the House Sparrow began to spread east from the Ural Mountains, following the growth of agriculture and, in particular, the building of the Trans-Siberian Railway, reaching the Pacific coast in 1929. It is steadily increasing its hold in Siberia and is now moving south into Turkestan following the route of the railway lines to Samarkand and Tashkent. These birds, however, unlike the ones that have spread from southern Asia are sedentary and as such are no doubt displacing the Tree Sparrows from the towns. I have been unable to find anything published since Carruthers's visit about the sparrows in this interesting zone of confrontation and it thus seemed an obvious place for a personal investigation.

My first attempt to put this into action was to visit the Intourist office in

London and enquire about trains from Moscow to Samarkand and from Moscow to Tashkent. This presented no difficulties – there are regular train services – but it was not until I said that I wanted to stop and spend a few days at each of the stations on the way that problems arose. The manager was summoned and when I explained what I wanted to do he was most sympathetic, but said this was not possible as there were no facilities for tourists at these places. My counter was that I would take a tent and sleeping bag, but he would not change his position and suggested that my best move was to contact the ornithological authorities (*sic*!) in the USSR and see if a joint expedition could be set up.

The ideal opportunity for this arose the following year (1970) at the four-yearly International Ornithological Congress that was to be held on this occasion in The Hague. My plan was to approach Academician G. P. Dementiev, the senior Soviet ornithologist attending the congress, to see if anything could be arranged. With this in view I prepared a short brief outlining the sparrow situation in Turkestan and suggesting it was of sufficient scientific interest to warrant setting up an expedition. Unfortunately, Dementiev was unable to attend the congress through illness (in fact he died a short time later) and the other Soviet representatives present did not show much interest in my project. I am unable to say whether this was because of concern about impending changes in the hierarchy of Soviet ornithology, knowing the situation about Dementiev, or because I failed to persuade them of the interest of the sparrow situation in Turkestan.

The next opportunity arose in 1975 when I attended a government dinner given at Lancaster House in London to make the presentation of the Tribology Gold Medal of the Institution of Mechanical Engineers to the eminent Soviet scientist Professor Igor Kragel'skii; this was to be received on

Boris A. Titov
SECOND SECRETARY
U.S.S.R. Embassy (Science dept.)
18 KENSINGTON PALACE GARDENS,
London, W.8

Boris Titov, the USSR Science Attaché.

his behalf by the Soviet ambassador. The ambassador was accompanied by the Science Attaché and it was with the latter I brought up the possibility of a sparrow expedition to Turkestan. He said that zoology was not his field of science, but if I would send him a brief of exactly what I was proposing he would pass it on to appropriate colleagues in the USSR to see what could be arranged. I updated the brief I had prepared for the 1970 Ornithological Congress and sent it off. Some weeks later I was speaking to a friend in the Foreign Office and, mentioning what I had done, asked if he thought there was any possibility of something coming of it. He asked the name of the Science Attaché and when I told him he laughed and said that the gentleman in question was not a scientist at all, but headed up the KGB in the London Embassy! I heard nothing further, but, recognizing that the prospects for the sort of trip to Turkestan that I had in mind were not good, decided that a visit to northern Afghanistan was the next best thing. This was before the Soviet Army had moved in! How I envied Carruthers his freedom of movement in this part of the world. (At the International Ornithological Congress held in Ottawa in 1986 I again approached the senior Soviet ornithologist, now Dr Vladimir Flint, about a trip to Turkestan and I still hope that something may be possible, but nothing has transpired so far.)

We flew to Tehran in May 1976 and, in addition to Afghanistan, took in parts of the Elburz Mountains, the southern Caspian littoral, the north edge of the Dasht-e-Kavir in Iran and a brief trip through the Khyber Pass to Peshawar in Pakistan. In Tehran we stayed at the Trog Hotel – at least that is what we called it from the only remaining letters of ARMSTRONG on the sign outside - and it seemed to fit. The Trog could scarcely be described as one of Tehran's great hotels; in fact it almost defied description and the less said about the loos the better. The food in Iran, however, was of a considerably higher standard than the accommodation. I grew quite fond of *chelam korush* – a tasty dish of chickpeas, meat and rice, described by my wife as yellow peril – that was a standard item on the menus of the little restaurants on the way.

In Afghanistan our route took us through Herat, Kandahar and Kabul, over the snow-covered Selang Pass at 3500 metres to the Amu-Dar'ya (the Oxus of ancient history) that forms the border between Afghanistan and the USSR, with a side trip to the remote Bamian Valley in the Hindu Kush. Presumably House Sparrows have to cross the Selang Pass on their way north to their breeding grounds in northern Afghanistan and Turkestan, less hazardous to them than for the Red Army whose convoys had to evade the Muhajideen ambushes there, but there is little for them in this bleak place and doubtless they will not pause. In the middle of May it was still covered with snow, country more suitable for Snow Finches, which were feeding on the ground beside melting patches of snow, than for even the hardiest House Sparrows.

Selang Pass, Afghanistan. A pretty inhospitable spot even in May.

This was an unforgettable experience in fascinating country and contact with a culture very remote from our own. I was particularly interested as an engineer to see the windmills near Herat. These differ from the ones we are familiar with in that the sails are mounted on a vertical shaft. This shaft is located in a cylindrical tower that has diametrically opposed slots cut out in the direction of the prevailing wind to produce the driving current of air. The vertical shaft can then be directly coupled to the grinding stone, thus obviating the use of the gear that is necessary with the European design with its horizontal sail shaft. By chance we found some schoolboys who knew about the windmills and were able to take us to them. These windmills were about 300 years old and had been in use until the beginning of this century.

It was very hot in May, even on the high Afghan plateau. We found the small squat cucumbers that were available on all the vegetable stalls to be very thirst quenching and to have the advantage that they could be peeled when we had no means of washing them. For drinks we stuck to the green tea that is available at the *chaikhanas*, or wayside teashops, that are to be found all along the road. Here one sits cross-legged on a dais, having removed one's shoes, and is served with a glass half-full of sugar and a small pot of green tea, costing about 2p. This is not only a very refreshing drink but, being made

Windmills near Herat, Afghanistan. The vertical shaft obviates the need for gearing that is used in European windmills.

from boiling water, seemed rather safer than the brilliantly coloured sweet syrups much beloved by the Muslim peoples and infinitely preferable to the American colas and other bottled horrors. This at least was the theory, but after succumbing to the local version of Delhi Belly, Montezuma's Revenge and Gippy Tummy that are the scourges of travellers in unfamiliar parts, I realized I had overlooked one vital fact. We discovered that in the Afghan towns the open drains that run through the streets also serve for washing the glasses. We also noticed that they were used to freshen up the tired salad vegetables in the market stalls and resolved not to eat uncooked vegetables! This, I am glad to say, is the only occasion that I have suffered from this unpleasant and demeaning upset. My wife unaccountably did not succumb. (Her weakness is in providing a desirable target for biting insects and thus she is a useful travelling companion as when she is around they tend to concentrate on her and ignore me!) This was a short sharp bout that did not take away from the excellent food we had – fresh nan, eaten hot within minutes of being taken out of the tandoori-type ovens, rich vegetable soups (well cooked!), *shishlik* (a lamb kebab) cooked over charcoal on the pavements outside the restaurants, followed by yoghurt or ice-cream and washed down with the inevitable green tea. As far as we could judge, the restaurant business is a highly specialized one. Each restaurant appeared to offer only one of these courses, but they were quite happy to send out a boy to a neighbouring establishment to bring back the other courses if asked, and all for less than 50p for the two of us. One other unfamiliar custom was that in the hotels one had to rent a bed, rather than a complete room, each room having from about six to eight beds. Being a Muslim country the other travellers were almost exclusively male, so that in these circumstances my wife stood out somewhat conspicuously.

There was not the problem I had imagined in communication, with Farsi the language of the western part of the country and Pushtu that of the east: English was rather surprisingly the *lingua franca* (is it now Russian?) even to the extent of appearing in the signs over the shops, despite the British having been thrown ignominiously out of the country on no less than three occasions. There were in fact few problems. On one occasion my wife restrained me from taking a photograph of a passing nomadic dromedary caravan, having seen at the same time a wife in the caravan restraining her husband from aiming his rifle at me. One has to be careful about photographing women in a Muslim country even with a telephoto lens!

My wife restrained me from photographing a passing caravan, having seen a wife restraining her husband from aiming his rifle at me.

Changing traveller's cheques in the provinces was also an interesting experience. The bank in Kunduz did not resemble a Western bank at all. We were ushered into a reception room with large easy chairs where we were hospitably entertained with the usual green tea while our cheques were being processed. From time to time an official came in requesting more information; it was all very formal. We were only allowed to change US $50 (about £25 at the time) each, whether because of some regulation or merely because Afghanis were in short supply in Kunduz we could not discover, and even then it took over an hour for the negotiations to be completed. However, £25 went a long way in Afghanistan in 1976 with everything extremely cheap for us.

Changing money in Tehran by contrast had provided no problems once one had established the bank hours. This we had not done in advance and arriving at a bank in the centre at 4.30 p.m. found it was not open until 5. We decided to wait outside. Within a few minutes the traffic was stopped and people began to line the sides of the street. We asked one of the numerous police officers what was going on, and he explained that the Shah would shortly be passing on his way to the airport to meet the Austrian Chancellor who was arriving for a State Visit. I asked if I could take a photograph of him when he passed. The people in front were moved away so that I could get a ringside view. It was evidently not considered a risk that I should point a 400 mm telephoto lens at the Shah from a distance of only a few metres. The cavalcade arrived shortly; the Shah's car was proceeded by about fifty motorcyclists and followed by an ambulance and a fleet of armoured cars, including one with anti-aircraft weapons. I was reminded of seeing Haile Selassie driving past in a limousine in Addis Ababa four years before. On that occasion there was one security guard beside the driver and a single motorcycle outrider. Neither strategy, however, was effective in preventing the fall of either monarch within a few years.

Frontier crossings were also very protracted affairs. The crossing into Pakistan had to be started early in the day as it was not permitted to enter the Khyber Pass after 3 pm. Just as in the days of the British Raj, only the roads in the North-West Frontier are controlled by the government; the rest is tribal territory and one is warned not to stray more than five metres from the road. This is not merely a formality; shortly after I returned home I read in the papers of two Americans who had been shot by tribesmen after they had ignored the warnings. A rifle and a bandolier seemed to be an essential part of the dress of the adult males. We visited the village of Darra in a remote valley; this is a centre for local arms manufacture. Here copies of internationally known weapons were being turned out on primitive pedal-operated lathes. I was surprised at first to find that the valley was completely devoid of birds, but the reason soon became clear – the intending purchaser wants to try out

Darra, North-West Frontier, Pakistan. One of the workshops where you can buy anything from a Colt to a Kalashnikov.

the weapon before clinching the deal. The valley resonated with shots and no doubt any bird that showed itself would have been quickly dispatched by these deadly sharpshooters.

However, let us return to the frontiers. One had to join separate queues to get emigration and currency clearance from the country of departure before going through health, emigration and custom checks from the country of entry. This process took about four hours – all the more galling as the transhumant pastoralists on their migrations from Pakistan into Afghanistan just wandered through without producing any papers. However, our passage was swift compared with that of three drivers of British lorries with loads of refrigerators for India. When we crossed into Pakistan they had already been there for two weeks. Since their last trip the regulations had been altered. Previously, the lorries had been sealed on entering Pakistan and provided the seals were still intact on leaving the country there was no problem. The new regulation required that duty be deposited on entering and reclaimed on leaving the country. This was beyond the means of the drivers and they were waiting until their employers in the UK could make the necessary arrangements. They were prepared for a long wait and were still there on our return into Afghanistan a few days later.

Another interesting feature at this frontier was the money changers on the Pakistan side. They had fibre suitcases stuffed full of different currencies and were quite willing to make any change you wanted – for all I know Hong Kong dollars to South African rand. All this was done mentally without resource to an electronic calculator or even an abacus, though at very favourable rates as far as they were concerned, judging by the fact that it was possible to bargain. What was even more surprising in that part of the world was that they seemed happy to walk off down the road apparently without concern, carrying the suitcases with their valuable contents.

The return crossing from Afghanistan to Iran was even worse, taking eight hours. The reason here was that everybody, and more particularly the vehicles, were being searched minutely for drugs. We saw the remains of a Danish caravan that had been found with drugs the previous day. A smart customs officer had spotted that one of the Formica panels had two new screws. After that, the caravan had been completely stripped to its shell; even the fuel tank and tyres had been cut open. No doubt this was a minor concern to the owners compared with the prospect of a long sojourn in an Iranian jail. To while away the long wait the authorities had provided a museum showing many examples of failed drug-smuggling attempts, such as boots with hollow heels and tins of fruit with concealed compartments inside. This was all the more ironic as one of the first sights on driving into Iran was the immense fields of opium poppies in full flower.

A joy in Afghanistan was the experience of being treated as travellers rather than as tourists, with the age-old questions: Are you married? How many children have you got? Where are you going? A charming meeting of this sort occurred one evening as we were walking back to our hotel along one of the main streets in Mazar-i-Sharif. A young man called from an open-fronted shop in which he was sitting to ask if we would take a cup of tea with him. It was a shop specifically selling tea, though oddly enough there was also soap on offer. A boy was sent for tea and cakes and we had a very pleasant half hour sitting talking to the proprietor. My wife wanted to buy some Afghan green tea and, during one of the interruptions when he was serving a customer, I suggested she should give him her custom. We got half a kilo of his best tea, but he would accept no payment, saying he had enjoyed the conversation too much for that. Abdul Samea had an ambition to go to America to further his studies. Like many Eastern countries Afghanistan then had a policy of sending some of its brighter students abroad for further education. This has not always had the desired result as many of the students have so liked conditions in the West that they have stayed rather than take their new skills back to their own country. According to Abdul, Afghanistan had an interesting solution to this problem. If the student overstayed his visit, his father was immediately clapped into jail and held there until the errant son returned: all right unless the son had fallen out with his father! We subsequently had an exchange of letters, when rather charmingly Abdul used to send his 'hot greetings', but this stopped when the Russians went ahead in the 'Great Game'. I wonder, is he still selling tea in Mazar-i-Sharif?

Mazar-i-Sharif was the closest to the Soviet frontier that we got to on normal roads; even then it was some 65 kilometres away with a large marshy area lying in between. There was a track to the border, but the bridge that formerly crossed the river had been removed to discourage unauthorized crossings – it has been reinstated since our visit to allow the Soviet troops to move south. Wanting to have a closer look, I asked a taxi driver to take me down. Apart from no doubt thinking this was a rather odd request, off we went, apparently without causing any concern to the authorities.

I made close observations on the sparrows wherever we went. In Tehran and Peshawar, at elevations of 1200 and 435 metres respectively, House Sparrows were well established and breeding, but above this, both in northern Iran and all over Afghanistan, which generally lies above 1200 metres, the House Sparrows were out in the open country away from human habitations and cultivation, nesting under bridges and in holes in the rocky embankments that had been cut out at the sides of the roads; in some places in Afghanistan they were apparently still on migration, no doubt on their way to Turkestan. In contrast, Tree Sparrows were present and well ahead with their breeding cycle in all the towns and villages.

On our way back we made a side trip to Bamian; this was particularly interesting. Bamian lies in a narrow valley with the fertile area restricted to the valley floor and the steep sides totally bare of vegetation. This valley was the last stronghold of Buddhism in Afghanistan, the natural defences supplemented by the commanding Red Fort at the entrance to the valley, resisted the Muslim invasion until the 8th century when it was overcome through the treachery of one of the princesses, who betrayed the source of water for the castle to the invaders. Bamian still possesses a natural difficulty of access; the road is very narrow. We were held up by a block between a bus and a gaily painted Afghan lorry with neither driver willing to give way. The lorry carried almost as many passengers as the bus and the impasse was finally solved by the combined passengers taking the initiative and physically manhandling the vehicles past each other with no loss of face by the drivers and only a minor loss of paint by the vehicles.

In Bamian itself there still remain two enormous effigies of the Buddha carved out of the cliff face overlooking the town – one is 53 and the other 37 metres high. The cliff is also honeycombed by cells that formed part of a great Buddhist monastery. At Bamian Tree Sparrows were nesting in the town, while the House Sparrows were confined to holes in the cliff face. Even at the isolated group of *chaikhanas* at the Band-i-Amir Lakes high in the mountains,

Traffic jam, Bamian Valley, Afghanistan. The passengers eased the vehicles clear, overcoming any loss of face by the drivers.

the nesting birds were Tree Sparrows and not House Sparrows. These lakes are very beautiful and very inviting in the harsh surrounding country; but one must remember that this is a fanatically Muslim country. Shortly after our visit a girl tourist was killed there by local people who objected to her displaying herself in a bikini.

In northern Iran and throughout Afghanistan wherever the Tree Sparrow occurred it was the house bird nesting in the towns and villages with the House Sparrow out in the open country away from habitations.

Willow Sparrows were also seen on a number of occasions in moist river valleys, such as the Bamian Valley, away from the towns, the sort of habitat in which I would expect to see them, but here they may be in conflict with House Sparrows, the two species arriving together as summer visitors from their wintering grounds in northern Pakistan. For example, in the valley of the Helmand near Kaleki there was a large collection of nests in conifers that from a distance looked just like a typical Willow Sparrow colony, but which on closer inspection turned out to be one of House Sparrows.

Three other sparrow species have been reported from Afghanistan. The Dead Sea Sparrow breeds in the great marsh of the Sistan on the border with

Habitat utilization by House and Tree Sparrows in Afghanistan.

Iran, where the Helmand River peters out into the desert. It has been reported north up the Farah Rud to where it is crossed by the road from Herat to Kandahar, but though I searched along the river in that area on both the outward and inward journeys I was unable to find any trace of it and had to wait until 1981 before I had my first encounter with this bird. The Cinnamon Sparrow occurs in the high country to the northwest of the Khyber Pass but, as I have already mentioned, it is unwise to go off the roads there searching and again I had to wait until 1981 before I first met up with this most attractive sparrow. Finally, the Saxaul Sparrow is believed to nest in the mountains at the junction of Iran, Afghanistan and the Soviet Union, but this lay to the north of our route and there was no time to explore that area. This was unfortunate as it is the only one of the twenty sparrows that has still eluded me.

He bowed . . . and passed on to deal with the goats on the nearby hillside that had been ignoring his shouted commands.

3: A biological mix-up: The Mediterranean basin

I remember the grave and learned Bevoriskius, in his commentary on the generations from Adam, very naturally breaks off in the middle of a note to give an account to the world of a couple of sparrows upon the out-edge of his window, which had incommoded him all the time he wrote, and at last had entirely taken him off his genealogy. 'Tis strange!' writes Bevoriskius, 'but the facts are certain, for I had the curiosity to mark them down one by one with my pen – but the cock sparrow, during the little time that I could have finished this note, has actually interrupted me with the reiteration of his caresses three-and-twenty times and a half. How merciful,' adds Bevoriskius, 'is heaven to his creatures.'

Laurence Sterne

The idea of a species is one of the simpler zoological concepts. It can be defined as a group of animals living in the wild that freely interbreed to produce viable and fertile young. That is not to say that hybridization between different species does not occur; the aviculturalist produces off-spring from some very unlikely crosses and natural hybrids certainly occur in the wild. The latter most probably arise when a member of one species strays into the range of another closely related species and, in the absence of a mate of its own kind, pairs with the next best thing. Certainly wild pairs of House and Tree Sparrows have been reported, producing offspring intermediate in plumage. These are clearly viable, though whether they are fertile is another question.

The concept of species is, however, a human one and whether birds conform to it is quite a different matter. A most peculiar situation occurs with the sparrows around the Mediterranean. In Spain, the Balearics, Greece and parts of North Africa, House and Willow Sparrows live side by side and behave properly as good species. In parts of Algeria and Tunisia, and to a lesser extent elsewhere, they interbreed freely producing a complete range of intermediates. Even more extraordinary, neither House nor Willow Sparrows occur in Italy, but a bird that is halfway between them and is difficult to place in either species. Cock House and Willow Sparrows look somewhat alike, the main differences being that the Willow Sparrow has a chestnut head compared with the grey crown of the House Sparrow, and a much more extensive black bib that spreads on to the flanks as prominent black streaks, whereas the underparts of the House Sparrow are unmarked pale grey. The Italian Sparrows have the chestnut crown of the Willow Sparrow, but the small bib of the House Sparrow and no flank streaks. The hens of all three birds are virtually indistinguishable from each other in the field.

Where the Italian Sparrow comes into contact with the House Sparrow there is a transition zone about 20 km wide, stretching from about San Remo in the west, roughly along the Italian side of the Alps, at an altitude of about 1000 to 1500 metres, to Trieste in the east, where the birds are intermediate in plumage between House and Italian Sparrows, whereas south of about Rome there is a more gradual transition from Italian to Willow Sparrows, so that in Malta the birds are almost, but not quite, pure Willow Sparrows.

The reason that I made several trips weaving across the frontiers between France, Switzerland, Austria and Italy, as already mentioned, was to plot the course of the transition zone in the north. This is a most attractive part of the world scenically and, quite apart from the interest in the sparrows, worth visiting in its own right. One thing that puzzled me at the western end was the lack of hotels in the Piedmont district of Italy. One drives across from the French side, where every village seems to have several hotels and one can have a lot of pleasure in speculating which has the best cuisine, to an equally attractive area in Italy where, away from the internationally known ski resorts, there appear to be few hotels and, at the time we were there outside the winter season, the few we found were closed for their holidays. The Latins, particularly the Italians and the Spaniards, in my experience appear to be highly urbanized peoples who regard the country as something for the peasants who have to live there, only occasionally to be visited for highly organized picnics, complete with the civilized appurtenances of tablecloths and wine coolers, or somewhere for the macho males to shoot at everything that moves, not a place to be enjoyed for its own sake. The beginning of *La*

Chasse in France is not a time for the timorous birdwatcher to skulk along behind hedgerows. Fortunately, I know of no birdwatchers who have come to a sticky end, though in their importunity frequent casualties appear to be inflicted by the hunters on each other.

In contrast to the problems in the Italian countryside, there is no difficulty in finding excellent facilities in the large urban centres with their cathedrals, museums and other ancient buildings to interest the tourist. I have not neglected those in northern Italy, where I have been to Florence, Pisa and Milan, though I must confess it was not primarily to look at the cultural artefacts, but to explore another peculiar sparrow situation that occurs there, as I shall describe later. My sympathies lie to some extent with the hot American lady, with her feet obviously hurting, whom I overheard speaking to her husband beside the Leaning Tower in Pisa: 'Hank, if there is no elevator in that tower, I'm not going up!'.

On French Corsica, the sparrows in consonance with the Italian place names, are pure Italian ones, whereas in Italian Sardinia across the narrow Strait of Bonifacio they are pure Willow (Spanish) ones. The birds have their separate rules. We did not go, however, to Sardinia via Corsica, but by ferry from the Italian mainland. The opportunity arose in September 1979 when I was halfway there attending a conference in Lyon. This conference was originally scheduled to end at midday on the Friday and I planned to drive from Lyon to Genoa to catch the ferry to Porto Torres, allowing for one night's stop on the way and the opportunity to check up on the transition zone from House to Italian Sparrows that I have just been describing. As it turned out, we did not get away from Lyon until the Saturday morning and it was not until that evening when we were having a pre-dinner drink in a hotel in Pinerolo some 200 kilometres from Genoa and my wife asked me what time we were due to board the ferry that I found it was in 15 minutes – I had completely forgotten to change the date of the ferry booking after I had found that the conference would not finish on the Friday afternoon. The following morning we made post-haste to Genoa, but as it was a Sunday the shipping office was closed until immediately before boarding time. I wondered how I was going to explain this situation in my limited Italian, but as so often in these circumstances the difficulty was overcome by a helpful stranger – in this case a multilingual Swiss – who was in front of me in the queue and handled the situation so well that we boarded the ferry without further incident just 24 hours late, helpfully muting my wife's comments on my organizational capabilities.

This was not the only problem on the way to Sardinia. Once we began our climb up the foothills of the Alps a nasty knock developed in the car as we went round sharp curves. A casual roadside look revealed nothing untoward and we proceeded, debating the cause of the increasing knock without even agreeing precisely where it was coming from. That evening in Pinerolo it rained so heavily that I shrank from doing any more than to check that all the wheels were tightly secured. The next day we were more taken up by the question of whether we would be able to get on board the ferry than by worry about what was beginning to sound a very expensive noise. That night the car was inaccessible to me between decks and it was not until the following afternoon when we stopped at a hotel in Bosa that I turned to it again. Once we had the car unloaded I began a full examination, but could find nothing amiss. On a road test the knock had gone. The only explanation I could come up with was that the cause was a loose empty wine bottle (an essential component of travel in the Mediterranean countries where they can be filled cheaply with the local vintage) in the boot. At any rate the car performed perfectly from then on.

It is said that problems always occur in threes. The third was one my wife had over the risk of being kidnapped, particularly when driving around Sardinia in a British-registered car, despite my assurance that our birdwatching appearance would dispel any thoughts of a worthwhile snatch. However, this was the year when Rolf Schild and his family had been kidnapped and his wife and daughter were still missing at the time of our visit. Before leaving home my wife confided to a friend that I was going to be quickly disillusioned if I had any thoughts about driving over the wild, sparsely inhabited country of the interior. From Porto Torres, where we left the ferry, we drove to Sassari, the second town on the island, along one of the flattest, dullest and dirtiest roads I have ever been on, on our way to the west coast. My wife was navigating and from Sassari she decided to depart from the main road to get some better scenery and within a few miles we were traversing some of the wildest country to be found on the island, or for that matter in Europe. That broke the ice and from then on we drove without inhibition all over the island, even stopping to have a drink in a small bar in Nuoro, the reputed headquarters of the *banditti*. The rather dour lady in black behind the bar looked as though she could have been in league with the local gang, but we survived and met with nothing but kindness wherever we went. On Corsica, on the other hand, I had an encounter with an obvious bandit with long moustachios and a fearsome look who I noticed was sitting outside his house when I was searching for birds around some unoccupied summer beach houses. He kept shouting at me, no doubt to warn me off the private property. At last he could stand it no longer and came over to remonstrate. When he got near, he bowed, wished me well and passed on to deal with the goats on the nearby hillside that had been ignoring his shouted commands.

Despite our initial impression on the way to Sassari, Sardinia, together with its neighbour Corsica, provides some of the most spectacular and exciting scenery to be met with in Europe. Driving in Corsica is quite an experience with a rock wall on one side and a precipitous drop on the other down to the sea or the bottom of a canyon; it is not for nothing described as white knuckle driving. I spent most of the time in second gear and we found it difficult to average as much as 20 km/h along the narrow and twisting roads. The regular sight of abandoned wrecked cars lying hundreds of feet below did much to concentrate the mind on driving and acted as a deterrent to bird spotting from the car. In complete contrast, the inland roads in Sardinia are wide, well engineered and surfaced, carrying very little traffic and apparently not going anywhere in particular. My wife had the scurrilous idea that perhaps the kidnapping and ransoming for which Sardinia is infamous was organized by the local government in connection with its road-building programme.

My main reason for visiting Sardinia was not to look at the Willow Sparrows, but a different species, the Tree Sparrow. At yet another engineering conference, this time in Naples in 1978, I was staying in a hotel on the front, though my bedroom window faced to the side. The morning after I arrived I went on to my balcony that overlooked the flat roof of the neighbouring building. Among other things the roof had a hen run and there, making use of the food provided for the hens, were, as might be expected, some sparrows. To my surprise, despite the fact that we were in the centre of the town, they turned out to be Tree Sparrows and not the Italian Sparrows that I would have expected. In fact, this was quite typical of the centre of Naples, where all the sparrows were Tree Sparrows and the Italian Sparrows were confined to the outskirts. Given half a chance, as I have already shown, the Tree Sparrow becomes a 'house' sparrow whenever the opportunity arises. Seeing the birds in Naples recalled a brief paragraph I had read in a book by Martin L. Cody. In this he quoted an observation by Hartmut Walter that 'on the island of Sardinia, *P. montanus* [Tree Sparrow] is restricted to the downtown and port area of the major city of Cagliari; the more typically urban sparrow *P. hispaniolensis* [Willow Sparrow] is displaced to the suburbs and countryside'.

On visiting Sardinia in November 1979, we found that Tree Sparrows were not only occupying the centre of Cagliari, as Hartmut Walter had observed, but were well established in a narrow belt along the east coast of the island, north as far as Olbia. In Cagliari, Tortoli and the neighbouring villages of Arbatax and Girasole, the Tree Sparrow was the dominant urban species with the Willow Sparrow confined to the surrounding countryside. In Olbia, and at Villagrande Strisaili and Villanova Strisaili, 15–29 kilometres inland, both species were present. It would appear that the Tree Sparrow is a comparatively recent arrival on the island and I suspect that they may have come to Cagliari by ship from Naples, where as I have mentioned it is the dominant sparrow. (This is not an unexceptional mode of travel by sparrows and has led to colonizations by different sparrow species elsewhere.) The Tree Sparrow now seems to be spreading from Cagliari and displacing the Willow Sparrow from the towns and villages as it goes, despite the fact that the latter is the larger bird. The Willow Sparrow is similar in size to the House Sparrow with a typical weight of 25–30 grams, compared to that of the Tree Sparrow of 20–25 grams, and though it will become a 'house' sparrow in the east in the absence of its more urbanized congener, the House Sparrow, it appears only secondarily at home in urban surroundings in the west. Certainly it was present throughout the rest of the island that had not yet been colonized by the Tree Sparrow in both the urban and countryside habitats and formed enormous noisy overnight roosts in trees in the squares of many of the villages.

The most spectacular of these roosts is in the Corso Vittorio Emmanuel in San Antioco on the Isla of that name that I found on a return visit to Sardinia in 1989. This street, about 200 metres long, is lined on both sides by trees that meet across the centre to form a shady tunnel. When I was there in November the birds started to arrive about 4.30 p.m. and by dusk the noise was deafening, a great waterfall of sound as each individual sought its favourite resting place in the trees for the night. In the morning the first birds were heard calling about a quarter to seven and most had departed in about half-an-hour, large flocks forming in the air above the roosting place and streaming away in all directions, even across the three kilometre open stretch of water towards San Giovanni Suergui on the mainland, disappearing from sight in the binoculars still high in the air.

There was no obvious way in which the number of birds occupying the roost could be counted, but judging from the volume of sound and the size of the flocks departing in the morning, there must have been several tens of thousands of birds present. The Corso is closed to motor traffic during the night and, before the street is blocked with other vehicles in the morning, the mechanical road sweepers are out by the time the birds have left, cleaning up both the roadway and the broad pavements from the droppings of the night.

Clearly such a large concentration must have drawn the birds from a large area, not only from the Isla di San Antioco, but also from the nearby mainland. Communal roosting on this scale must obviously be a considerable advantage to the birds to compensate for the energy expended in flying from such distances. Most probably it serves for information exchange about good feeding areas in the surrounding agricultural land, allowing maximum benefit to be obtained from varying and unpredictable sources of food.

In the 10 years between these two searches on Sardinia the sparrow position had changed dramatically with the Tree Sparrow now present in all the towns and villages around the coast, not confined to a narrow band on the east coast as we had found in 1979.

This turned back my thoughts to the position of the sparrows in Naples where the Tree Sparrows were more urbanized than the Italian ones. What were the latter birds? Were they closer to House or to Willow Sparrows? This led to a tour of the towns of northern Italy where Italian Sparrows are at their most typical – those from Rome southwards showing more and more Willow Sparrow influence. The question, as with many simple biological ones, did not yield a straightforward answer. Both species seemed equally at home in the town centres and either could have justified the name 'house' sparrow.

Birds of typical Italian Sparrow appearance are also to be found on Crete and here they fully justify the appellation 'house' sparrow. As in so much of the Mediterranean area there is a major building programme on Crete with new hotels and apartment blocks. The partly finished buildings, still awaiting their final rendering and plastering, provide an almost continuous supply of crevices and holes that give suitable nesting opportunities to be exploited by the local sparrows. I saw both Tree and Willow Sparrows when I was on Crete in October 1980, but these were out of the towns in the country districts and I suspect they were passing migrants.

The sparrow situation in northeastern Algeria and Tunisia is rather different. Here we do not have a stable population intermediate in plumage like the one in Italy, but rather a complete range of hybrids going from birds at one extreme that look like pure House Sparrows to those at the other that cannot be distinguished from pure Willow Sparrows.

While the Latin peoples can be described as largely indifferent to bird-watchers, the same cannot be said of the North African Arab children. Whenever one tried to look at birds, they arrived in hordes and jumped up and down in front of one's binoculars or camera screaming loudly; by that time it did not matter anyway as the birds had already fled. That is when you do not want help. When you do it is a different story. We were driving along in the hinterland one day after some heavy rain when the rented car came to a halt with the gap between the wheels and the wheel arches completely jammed with glutinous mud; I have never encountered more insidious and tenacious material anywhere. This was in a semi-desert region, there was no one to help us and not even a branch to lay on the road or a stick to poke at the mud. It was a laborious job to remove enough with our hands to free the wheels so that we could make a few metres progress before grinding once again to a halt. It was some hours before we finally got on to a hard surface and by this time we and the car were completely filthy, both inside and out. When I returned the car to the rental company a few days later I carefully parked it round the back. As I completed the necessary documentation I remarked to the clerk that the car was in need of a bit of a clean. Fortunately, he did not ask to see it, but merely replied that they always came back a little

Whenever one tried to look at the birds they arrived in hordes and jumped up and down screaming loudly.

dirty and they gave them a good clean before renting them out to the next customer. I wished him well and beat a hasty retreat.

The mix-up of sparrows around the Mediterranean basin is a fascinating biological puzzle that has attracted the attention of a number of eminent ornithologists from Professor Wilhelm Meise of Hamburg, who published his findings in 1936, to my good friend Professor Richard Johnston of the University of Kansas who reinvestigated the position in 1969. Apart from the areas already described there is also a tendency for the House and Willow Sparrows to hybridize in other places in the Mediterranean area; for example, J. K. Stanford reported hybrids from Cyrenaica in Libya in 1954 and in 1975 Portuguese workers reported hybrids from central Portugal. In Yugoslavia I found a small hybrid population at Plavnica beside Lake Scutari, though at Titograd only 15 kilometres away the two birds were behaving normally as separate species, living together side by side with no sign of interbreeding. We found this colony by accident and as it was getting late in the afternoon it was obviously worthwhile staying to have a more detailed look. We had just passed through the village of Virpazar where I had noticed a small inn. There I asked for a room in German, but no one seemed to understand me and even my parrot-learned *Imate li soba?* ('Have you a room?' in Serbo-Croat) did not help. After a short time the chef was produced, but his German was even worse than mine and we seemed to be getting nowhere until I happened to mention that we came from England. The atmosphere changed. His German at once became fluent and in a moment our cases were being collected from the car and taken up to a room – something rather exceptional in communist Yugoslavia where you were expected to carry them up yourselves. This was in the early 1960s when memories of the German occupation still ran high and although many people spoke German they did not like to acknowledge the fact to German tourists. At least I was complimented by the fact that they took my German as coming from a native speaker.

The occurrence of hybrid populations is a very unusual phenomenon in the bird world. Obviously, the House and Willow Sparrows are very closely related and, while they make things rather untidy for the precisely minded systematist, their unusual interrelationships round the Mediterranean in fact provide a useful clue for the unravelling of the evolution of the sparrows.

. . . according to the operator the problem was that the monkeys had pulled the lines down.

4: Sparrow pots: Melita, Cossyra and Serendip

Another story make from wast to chinne,
with breasts like pots to nest young sparrowes in.
Thomas Randolph, '*To his well Timbred Mistresse*', 1640

My attention was first drawn to the term 'sparrow pot' in 1953 in a short note by the German ornithologist Dr Wilhelm Meise in the magazine *British Birds*. This gave a brief account of these unglazed earthenware pots that were hung up on house and barn walls to provide nesting places for sparrows – as precursors of the modern nestbox. Meise suggested that they probably originated in Delft in The Netherlands in the 16th century. Delft was an important brewing town and the sparrows attracted by the grain were recognized as a great pest, the pots being put up to allow them to be caught and easily controlled. Professor Maury Meikeljohn, however, later drew attention to the presence of pots in a painting by the Flemish artist Pieter Breughel that dates from the 1560s. This confirms the general area for the origin of the pots as the Low Countries, possibly early in the 16th century. It seems most probable that sparrow pots were introduced to England early in the 17th century by the Dutch engineers who were involved in the draining of the Fens.

Collingwood Ingram describes a *pot de moineau* that he obtained in the First World War in the Vosges Mountains, where he said they were still common at that time on farmhouses. This was a rough hemisphere, about 15 cm in diameter with a short projecting spout to provide an entrance for the birds. The pot was hung on a nail so that the flat back rested against the wall. The back was left open so that the pot could be taken down and the nestlings removed before they fledged. In addition to giving the farmer a means of controlling the sparrows that were considered as a pest, the young birds provided a supplement to the family diet, ending up in a different pot to be converted to sparrow pie. The sparrow pots also provided an alternative site to the thatched roof where they could cause considerable damage; quite apart from the deleterious effect this can have on the integrity of the thatch, there are a number of reports of sparrows actually causing fires by carrying smouldering straws to their nests in thatched roofs.

There is little information on how common or widespread the use of pots was in England, though published records seem to be limited to East Anglia and the southeastern counties of Kent and Sussex. It appears that they were in regular use in Kent up to the middle of the 19th century. By the beginning of the 20th century, however, they had become little more than a curiosity and it was difficult to find unbroken specimens. In consequence I had little hope of actually seeing one. However, C. B. Ticehurst, visiting Malta in 1911, described earthenware bottles hung up under the eaves of houses there to provide nesting places for sparrows. In correspondence in 1961, Dr C. de Lucca was able to tell me that these nesting bottles were still in use, though fast disappearing. Ticehurst described them as long earthenware bottles with the bottoms knocked out and Dr Lucca confirmed that they were in fact imported earthenware gin bottles giving something very similar to the pots that had been used in the Low Countries, England and France. He said that the object of putting up these bottles was to allow the nestlings to be removed just before fledging so that they could be hand reared and satisfy the local craving for cagebirds. For those who did not want to rear the birds themselves, it was said that if the cage was hung up beside the nesting bottle the parents would continue to feed their young through the bars. Dr de Lucca told me that the practice was dying out as this type of bottle was no longer imported and the existing ones were disappearing through breakage.

It seems doubtful if there was any connection between these bottles and the pots that originated in the Low Countries, apart from the fact that they probably originally contained Dutch gin, but nevertheless I was curious to see one. It was not, however, until October 1977 that I was able to visit the

Willow Sparrow nest, Malta. A substitute for the earthenware gin bottles previously used.

island, by which time Dr de Lucca had died. I turned to Joe Sultana, President of the Malta Ornithological Society, and to my delight he knew of one nesting bottle still in use. At the first opportunity we rushed off to see it. When we arrived at the spot he told me to drive slowly. He appeared puzzled; after a search he decided that it had gone and in its place there was a section of pipe for carrying underground electric cables. This was no doubt just as attractive to the sparrows, and indeed there was a nest present, but it was not what I had come to see.

This was a great disappointment and it seemed that my last opportunity had gone. Imagine my surprise, when going to stay with my friend Dr Upen de Zylva in the provincial town of Kurunegala in Sri Lanka in 1979, the first thing I noticed was a sparrow pot hanging up on the wall of his house. Sri Lanka is a Buddhist country and I could not believe that it was there for the purpose of taking young sparrows. No! Upen tells me that the pots are put up because it is considered lucky to have sparrows about the house. It is even possible that the householder could gain merit by doing a good turn to an animal. This brought to mind a situation I had seen in a Buddhist temple in Bangkok. In the temple grounds there was a lady sitting by a number of small cages with birds, including sparrows, in them and a notice in English 'PLEASE SET FREE THESE POOR BIRDS. You will be happy and prosperous.' I could see that I could possibly gain merit in this way, but what about the trappers? Were they building up a stock of demerits?

Upen de Zylva tells me that the pots are still commonly put up on the walls of homes in the villages and those of middle-class houses in the suburbs in Sri Lanka. Both cooking pots, or *chatties*, and water pots with a jug like neck,

Sign in Bangkok temple. Do the birdcatchers get demerits?

rather than the specially made pots that were peculiar to Europe, are used. The neck is placed against the wall and a hole made in the bottom to give access to the birds. These pots are usually daubed with 'Chunam', a white calcium paste that is used with betel leaf for chewing, as a deterrent to predators. The history of these pots is obscure, but it is thought that they predate the arrival of Westerners in Sri Lanka and thus originated quite independently of the European ones.

Thus my story had a happy ending and even more so in 1981 when I visited another sparrow enthusiast, Drs Minouk van der Plas-Haarsma, in Amsterdam and she most kindly presented me with a sparrow pot: not an original, but a modern replica. The final twist is that although this pot has hung on the wall of my house ever since then it has so far failed to induce a pair of sparrows to nest in it, although a young pair showed interest in the pot, even taking in nesting material, from November 1983 until April 1984. At the beginning of the breeding season, however, they apparently lost interest and disappeared. Sparrows had ceased to breed in my garden a few years before

Modern Dutch replica of sparrow pot. Hung up on my house it has failed to attract buyers.

and even this desirable home failed to induce them to come back. It still hangs there as a reminder of my search for pots, only occasionally attracting the attention of passing tits and Starlings.

Although Joe Sultana failed to turn up a 'sparrow bottle', he nevertheless did much to make our stay on Malta both rewarding and enjoyable, though he did his best to nip a promising friendship early in the bud. During our visit there was a meeting of the Ornithological Society at which some films were being shown. We were invited to come along and after the films Joe introduced me as the author of a book on the House Sparrow and suggested that as there were still 20 or 30 minutes before the end of the meeting perhaps I should like to say something about the sparrows on Malta. After his kind words I could do nothing else, breaking the golden rule that one should never get up in front of an audience to speak without having made some preparation beforehand. It says much for Joe's pleasant personality, to say nothing of the unstinting help that he gave me, that I have forgiven him for the embarrassing situation that he dropped me into.

Malta should be a paradise for birdwatchers, lying as it does on an important migration route between Europe and North Africa. Much of this is spoiled, however, by the passionate mania the Maltese have for shooting birds. And not only birds: when there are no migrant birds passing through, I have actually seen them shooting butterflies! Joe Sultana and his colleagues in the Ornithological Society have embarked on a battle to save the birds, establishing reserves and undertaking an extensive educational campaign in

Sticker used by the Ornithological Society, Malta, in its efforts to reduce the bird carnage on the islands. It reads "I came, I made you happy; let me live free".

the hope that the next generation will grow up with a respect for birds and that Malta can take its rightful place as a haven for birds and birdwatchers and gain the foreign currency the latter bring in. Everyone interested in natural history must wish the Ornithological Society every success in its endeavours.

Both Meise and Johnston placed the sparrows on Malta as intermediate between Italian and Willow Sparrows. I was interested to discover whether in their habits there was any evidence of House Sparrow blood. The situation turned out to be most curious. On the east side of the island, where most of the human inhabitants live, the sparrows, despite their marked Willow appearance, were behaving almost exactly like House Sparrows, living in the towns and villages and nesting on the houses; in Valletta I even saw one of the group displays in which a number of males pay court to a single female that is so typical of the House Sparrow, but which otherwise I have never seen in a pure Willow Sparrow population. In contrast, the small parties of sparrows that I found on the uninhabited, uncultivated, rocky western side of the island were much less approachable and were behaving far more like genuine Willow Sparrows. This raises all sorts of questions: do the 'town' and 'country' birds maintain a separate existence with the eventual possibility of evolutionary separation, or is there interchange between the populations? This question is one that can only be answered by individual colour-marking of the birds, but would be an interesting study if some Maltese ornithologist could be persuaded to spend some time with sparrows rather than with the more 'exciting' birds that pass through the islands.

With the biologically confusing situation in Tunisia and my doubts about the purity of the Willow Sparrows in Malta, a visit to Pantelleria, ancient Cossyra, lying midway between Sicily and Tunisia seemed worth a search. This we carried out in 1988, but it did little to clear my confusion.

Pantelleria is a fascinating place. I intended to rent a car on arriving at the airport, but Pantelleria is not like that. However, while I was arranging seats on the bus into town my wife had engaged an attractive Italian in conversation and he put us on to a garage where we picked up a rather dubious Fiat 500. This was a Friday and I had not even enough money to pay a deposit. I asked if he would wait until the Monday when the banks opened. The owner was not worried; he had no need to be in Pantelleria – Pantelleria is a tiny volcanic pimple in the Mediterranean, 50 km round, and there was no way to escape except via the airport. Neither was he worried about driving licence or insurance. His main concern was that I should not forget to keep the engine topped up with oil: two litres when we picked up the car and regular additions thereafter. It did not let us down; the only difficulty was that after a couple of days the driver's door refused to open and the driving seat had to be approached via the passenger's side – no real problem to Britishers used to entering cars from the right-hand side.

In the Hotel del Porto, where we stayed, sparrows were nesting in a hole on the balcony of our room – the only balcony so occupied. The only other occupants – of the hotel, not our balcony – were two British engineers from the Italian company that had installed a desalination plant with a capacity of 60 tonnes/day, but that as yet had not produced more than 5% of the designed output. It is the only hotel I have stayed in where the water from the taps appeared to come directly from the sea: too salty even for washing teeth! What it was going to be like in July, when the summer invasion of Italian holidaymakers would be at its height, I dreaded to think. This was a plumbing problem even beyond the skills of my wife (see Chapter 5).

The local people were very friendly and charming, but very insular. They would not drink the local wine because they feared the effect of a fungicide that was sprayed on the vines and turned the leaves a bright sickly green. They preferred the imported Sicilian wines, apparently unaware, as we had noticed the previous week, that exactly the same fungicide was in use there!

Sri Lanka, my other 'sparrow pot' location, could not be more of a contrast to Malta with the deep respect that the Buddhist population has for its animals. Lying between 6° and 10° north of the equator, it is the best place to visit if one wants to see House Sparrows in a tropical situation, as it forms the southernmost limit of the bird's natural range. The term 'natural' must be stressed in this context, since as a result of introductions that occurred from the second half of the 19th century onwards the House Sparrow now occurs closer to the equator through its spread in the southern hemisphere, for example, in Mombasa in Kenya (4°S) and Salinopolis in Brazil (0.5°S). On the west coast of South America it even crossed the equator in Ecuador in 1977 and has now spread to Buenaventura in Colombia, nearly 4°N. Sri Lanka, where the bird occurs at Matara, 6°N, still remains the closest approach in the northern hemisphere, although this is being threatened as the birds introduced to North America are now spreading through Central America and have reached Concepción in western Panama, at 8.5°N.

We made our first trip to Sri Lanka in February 1975, visiting the east coast from Colombo north to the Wilpattu National Park and inland to Kandy and the tea-growing hill country of Hunasgeriya. Most important was our meeting with Dr Upen de Zylva, the eminent Sri Lankan bird photographer. I had been in touch with Upen before we left England and he suggested that I should telephone him as soon as we arrived. Hunasgeriya was not far from his home in Kurunegala, but it was not easy to make contact. The first evening I failed to get through – according to the operator the problem was that the monkeys had pulled the lines down (British Telecom have not thought about that one yet!). It was the same story the following evening, but on the third attempt I did at last manage to get through. The monkeys had obviously been having a field day.

Our meeting resulted in a pressing invitation for a return visit and a much more extensive exploration of this beautiful island; this we did in February 1979, when, in addition to the parts previously covered, we also visited Yala National Park in the south, the one remaining area of tropical rain forest at Sinharaja, another area of tea-growing hill country at Lindara, and Nurawa Eliya and Horton Plains on the roof of the island at 2000 metres. This allowed us to see something of the three major climatic zones – the southwestern low country wet zone, the dry low country zone at Wilpattu and Yala, and the hill country. The contrast is remarkable, going from humid tropical heat on the coast to the bracing climate of Horton Plains with frost at night. This was quite a shock for our driver who had never been there before and was quite unprepared for freezing conditions.

We used Colombo as a base from which we made trips to other parts of the island, staying with Mrs Chandra Dissanayake. She is a very talented cook and has published a book on Ceylon cooking. Dishes in Sri Lanka are spicy and I suppose could be covered generally by the English term 'curry', though they are much more subtle than this implies, each dish with its own blend of spices ranging from the hot to the very hot and generally increasing in spiciness from breakfast through lunch to dinner. There was always the accompaniment of tea. The people are naturally very proud of their tea with its international reputation and are most punctilious about its freshness and quality. I rather lost face as I am not a tea drinker. We went round the factory on one of the tea estates on which we were staying and after sampling the different blends were asked to say which we liked best. I thought that my choice of the poor quality big leaf tea that could only be sold to the undiscriminating Arabs would lose me respect until I found that the owner himself did not like tea, preferring the local *arak*. I went along with him in this.

One thing that puzzled me about Sri Lanka was the cheese or rather the lack of it. I am fond of cheese; when visiting strange parts I go out of my way to sample the local varieties and on occasion have enjoyed sheep's, mare's and camel's milk cheeses in addition to the more familiar products from cow's and goat's milk. In Sri Lanka, however, we saw only imported cheese and wondered if an indigenous cheese was lacking.

I have driven in most of the countries we have visited with two exceptions: India and Sri Lanka. In both these countries the roads are crowded with every conceivable kind of traffic and even if this means that one has to drive slowly the risks of accidents are still high. As a European one does not want an accident where there is a possible chance of nationalistic feeling running a bit high. Driving in Sri Lanka is particularly exciting. As a heritage of British rule, driving is on the left, but as a practical expedient everyone seems to drive more on the right to avoid the slow moving bullock carts and pedestrians at the side of the road. This means that approaching vehicles are to one's left; that is until the last moment when both vehicles veer sharply to the left and weave past as required by the traffic regulations. This is quite unnerving until one gets used to it. I do not know if there is any religious connection, but there seems to be a similar fatalistic view on driving to that one meets in Muslim countries, where, if accidents occur they are the result of providence, or lack of it, on the part of the Almighty rather than the responsibility of the driver. Without being presumptuous I prefer to do all I can to assist personally in these matters.

We went to Yala in two vehicles, one driven by Upen de Zylva and the other by a friend of his, P. P. Piya Jinadasa. Piya is a manufacturer of tea handling equipment and the way from Colombo to Yala passed his factory, where we stopped to have lunch with the manager. We were planning to stay in a bungalow in the Park and regulations do not permit entry after 5 pm. With the excellent but protracted lunch, the traffic on the road, and one or two other calls, it was clear that we were not going to meet the deadline. Piya with his characteristic resource remembered he had a friend who was the governor of a local jail and there we went for accommodation. Some of the trusties fixed up our cells with beds and we spent a comfortable night. Not fancying the prisoners' fare, we repaired to a local restaurant. Having fed well we paused on our way out of the grounds to admire a half-grown elephant

As we were standing close-by it whipped out its trunk and snatched my wife's handbag.

that had been hand-reared after its mother had been killed. As we were standing close by it whipped out its trunk and snatched my wife's handbag. This – the handbag not my wife – was carefully held down on the ground with one foot and opened with the trunk. Each item was carefully removed and examined individually. We were somewhat concerned as the bag contained both my wife's passport and traveller's cheques. Some of the restaurant boys came to our rescue by offering it more attractive food and we recovered the bag. Apart from the bag itself, which was slightly deformed and would no longer shut properly, and lipstick that the elephant had rather fancied – in the excitement I failed to notice its sex – nothing was damaged, not even the small mirror that had been extracted with the other items, though my wife's sun hat has never looked the same!

Another guest of the prison governor that night was Laddie Outschoorn who had played cricket for Worcestershire in the 1930s. It was interesting to listen to his memories of cricket at that time. He had come just too early on the international cricket scene, when even successful 'players' did not make much out of the game and before Sri Lanka was rightfully acknowledged as a first-class cricketing country.

Wilpattu provided another interesting experience, rather different in character. I am no photographer, but I do like to take the odd picture of the exotic birds to incite jealousy in my birdwatching friends back home. My technique is to take a quick long-range shot so that at least I have a record and then try to move closer to get something better. At one of the tanks in the Park where we were allowed to dismount from the vehicles, I got my first sight of a Painted Stork. I had just taken my first exposure when there was a shout from the ranger to get back on board. In this situation one does not pause; he had heard the monkeys alarming and we were no sooner safely back in the Land Rover than a magnificent leopard walked up and lay down on its back with its legs in the air just beside us. I failed to get my closer shot of a Painted Stork, but instead managed some reasonable pictures of a leopard with a 400 mm lens, even though these had to be taken at 1/30 second as the light was going. So much for the photographic pundits who say that it is not worth using a hand-held telephoto lens at slower speeds than 1/250 second.

Upen de Zylva, in addition to his superb achievements as a bird photographer, is no mean naturalist; so it was somewhat ironic that we should be able after only 10 days on the island to introduce to him a new bird species. From Kurunegala we went to spend a few days at Hendala on the coast north of Colombo. When we had been there for a day Upen telephoned to see how we were getting on. I casually remarked that we had found no less than four species of kingfisher in a small creek near the hotel in which we were staying. Three were to be expected – Eurasian, White-breasted and Stork-billed – but the fourth, the Black-capped Purple Kingfisher, which is listed in the most recent check list of the birds of Sri Lanka as 'a rare winter vagrant in the low country', was something of a surprise to Upen who had never seen one. He was down at dawn the next morning and I was pleased that he found it in the same place and was able to confirm my identification.

I found the House Sparrow much as described in Phillips's check-list 'locally distributed, but generally plentiful near human habitations throughout the lowlands and in the hills to altitudes of over 6000 ft'. My limited observations, however, gave me the impression that the bird appears to have a lower density than in comparable regions at higher latitudes, an impression I have since confirmed in other tropical areas, suggesting that they form a marginal region for the House Sparrow. This was true even in the more temperate tea plantations and at Nurawa Eliya in the hill country. To my surprise they were, however, present at the isolated office buildings in the rain forest at Sinharaja. These offices were built as part of a project to exploit the last-remaining rain forest for plywood manufacture. It was a far-seeing move of the newly elected government to abort the plywood project and declare Sinharaja as a nature reserve. The office buildings are now available for visiting naturalists to stay in, though as there is no telephone it is a little difficult to make arrangements. That was all handled expeditiously for us by our Sri Lankan hosts, on this occasion Nanda Senanayake and his wife Mano. We had been warned that it would be hard at night on the concrete floor, but our luck was in when the manager told us that the beds ordered seven months ago for visitors such as ourselves had arrived that very day. We were the privileged occupants to handsel them. With the timber extraction roads making the forest accessible, Sinharaja is a fine place to see rain-forest species and some of the 21 endemic species of birds in Sri Lanka, which are mostly to be found in the hill country. The rain forest lived up to its name during our short visit; it poured, but we did manage to see the magnificant Ceylon Blue Magpie none the less. Whatever difficulty we had in locating these birds was not shared by the leeches that abounded in the forest in finding us, even when we picked our way circumspectly in the centre of the tracks away from any vegetation. They were very effectively discouraged from overstaying their welcome by the application of a piece of soap.

One very noticeable difference in the House Sparrows in Sri Lanka was that they were even more of a 'house' sparrow than with us, as they came freely inside buildings. At the isolated bungalow in the Yala National Park where we stayed, the House Sparrows joined us for our meals in the living room, perching on the wall lamps and on the backs of chairs; at night a pair came into our bedroom to roost on the bracket supporting the fan above our bed. They were also seen frequently entering the open-fronted shops that are a common feature of the East, plundering the open sacks of grain despite the best attentions of the owners.

A small number of House Sparrows were liberated at Lajes Airport on Terceira in 1960.

5: The colonists: The Maraconese

Who wonder 'mid our fen why men depart
To seek the Happy Isles.
Rudyard Kipling, '*Song of the Cities*': Auckland, 1893

Before you reach for your gazetteers, let me explain that the Maraconese is a rather obscure name for the four groups of volcanic islands in the eastern Atlantic: the Azores, Madeira and Porto Santo, the Canaries and the Cape Verdes. Sparrows occur on all of them: House Sparrows were introduced to the Azores in 1960; Willow Sparrows arrived on Madeira in 1935; the same species is found throughout the Canaries, having first been recorded in 1856. No less than three sparrows occur on the Cape Verdes: the endemic Iago Sparrow and the Willow Sparrow that Charles Darwin found when he visited the archipelago in 1832 and the omnipresent House Sparrow, recorded for the first time in 1924.

We went to the Azores in May 1984. In addition to feeling that she has to accompany me on my travels to avoid my getting into difficulties, my wife has a mission abroad – to improve the plumbing for those 'lesser breeds without the law'. Her record in this field is impressive and nowhere better than in the Azores. In Horta, the well known stopping place for trans-

Atlantic yachtsmen and the principal town in Faial, we stayed at the Estalagem Santa Cruz, built originally as a fort in the 16th century. It is a most luxurious hotel, but there was one major defect: the basin in our bedroom had one of those plugs operated by a rod between the taps. Or at least that is the principle; on lifting this rod it came away in your hand. That was a major problem as one of her first tasks on arriving at a new destination is to wash a few minor items – a difficult proposition when you cannot retain water in the basin, compounded by the fact that the squash ball taken for such emergencies could not be used as the ineffective stopper got in the way. The management were informed and to their credit a complete new basin together with a working plug were in place when we returned that evening. Our next stop was on Terceira. Here we found a small motel near the airport where we had a suite – sitting-room, two bedrooms and, to my wife's joy, a kitchen complete with a sink and hot water geyser. We had scarcely had time to unpack before the sink was filled with the previous day's cast-off clothing. Everything was fine until my wife removed the plug and found herself standing in a couple of inches of soapy water – the outlet from the sink had no connection to the drain. It was a Saturday evening and the manager held out little hope, but that night at dinner he rushed in with a broad smile to tell us that all was well. A complete new plastic pipe had been fitted. The thought that the pleasure these two simple improvements would give to future visitors to the Azores warmed my wife's heart.

A small number of House Sparrows were liberated at Lajes Airport on Terceira in 1960 and the reason I went was to check up on how they had fared. The Azores have, however, a much earlier claim to ornithological fame: the Portuguese word *açor*, from which the islands take their name, means Goshawk. The Goshawk does not occur; the Portuguese who first discovered the islands misidentified the Buzzards that were said to have swarmed. They are still there, but the more recently arriving sparrows have done much better. When we were there in 1984, 24 years after their introduction, they had already spread to seven of the nine inhabited islands in the archipelago (they are now also on the two remaining islands) and I estimated that the total population had grown to about 50,000 birds. To my surprise I found that Dr Gérald le Grand and his colleague Maria de Fátima Melo of the Zoology Department of the University of the Azores in Ponta Delgada on São Miguel were making a special study of the sparrows on the islands. I am most grateful to them for supplementing much of the information I was able to obtain on my short visit. No doubt the sparrows have spread by hitching a lift on the grain ships that sail from island to island, but Gérard le Grand tells me that, in keeping with their successful profile, they have now been recorded using the inter-island planes.

The Azores have a very pleasant climate, mild enough for grass to grow all

The classical cone of Pico in the Azores.

the year round with the cattle remaining out in the fields, even being milked there in portable, diesel-engine driven milking parlours, yet warm enough to have vineyards and grow some excellent wines. Perhaps the most striking feature was the abundance of flowers set off by the hydrangea hedgerows that bordered the fields and were just coming into flower at the time of our visit in May. A more peaceful place, far away from all the hassle of modern living, would be hard to find. However, the volcanic eruption of Capelhinos that occurred in 1957 and lasted for two years, adding initially some 2.4 square kilometres to the island of Faial and the earthquake damage still evident to the buildings in Angra do Heroisme on Terceira are a reminder of the volcanic origin of the archipelago and a recognition that this phase is not yet entirely past. Pico, with its classical cone reaching a height of 2351 metres, is a weird island, much of it almost bare tufa with the vines growing in tiny 'fields' a metre or so square hollowed out of the rock and each containing one or two vines. These seemed to do well in the volcanic soil to produce the famed *verdelho* wine.

Willow Sparrows were unknown in Madeira until May 1935, when following a period of nine days of continuous strong easterly winds they turned up at various places and settled down. Madeira is 600 kilometres from the nearest part of the African mainland occupied by Willow Sparrows, but this bird is one of the most mobile of the sparrows, a regular migrant in parts of its range and a noted wanderer in North Africa. There are earlier records of sparrows – said to have been House Sparrows – on Madeira, dating back to the 19th century and including a small flock that remained on the island of Porto Santo from 20 January until 26 April 1903. It is difficult to account for the origin of these birds unless they came from a ship, as not only is the House

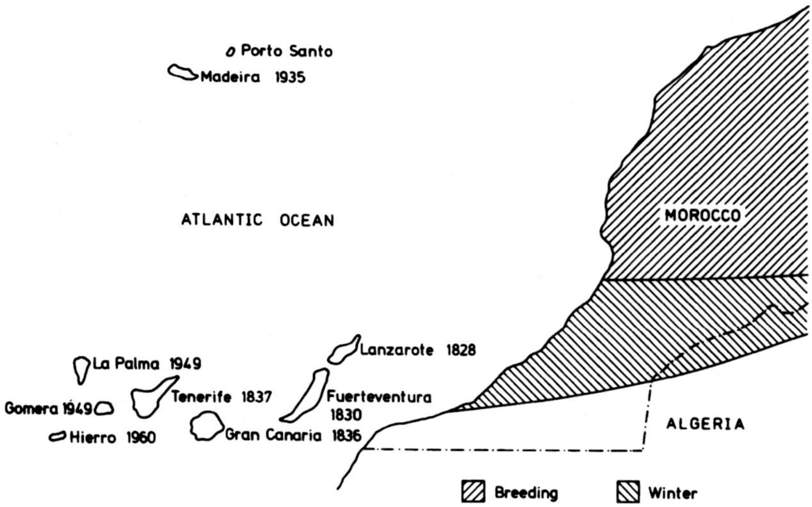

Distribution of Willow Sparrow in north west Africa and first recorded dates on Madeira and the Canaries.

Sparrow much more sedentary than the Willow Sparrow, but also the nearest part of its range lies much further away. Their subsequent disappearance is equally baffling, for the House Sparrow is a conspicuously successful colonist elsewhere, capable of building up a population from a small initial stock as we have seen in the Azores. However, since my visit in 1976 they have apparently recolonized, as a correspondent, Volker Konrad, found sparrows with grey heads in Funchal and Canico in 1980 and again in Funchal in 1984.

The history of the Willow Sparrow on the Canaries is rather obscure. Popular opinion is that it was introduced. The first-recorded date is 1828 for Lanzarote and if those for the different islands are examined, there appears to be a gradual spread from east to west. This is very reminiscent of the known spread of the House Sparrow on the Azores from its point of introduction on Terceira. Lanzarote, being one of the more sparsely inhabited islands, seems an unlikely place for an introduction to have been made; moreover, it lies only some 200 kilometres from the nearest wintering place of the bird on the African mainland. If this species arrived naturally on Madeira, distant 600 kilometres, it seems likely that it could have readily reached the Canaries under its own steam. The earlier colonization of the Canaries, much closer to the African mainland population of Willow Sparrows, is not surprising.

In order for colonization to succeed, not only must sufficient individuals have arrived to provide the necessary gene pool to allow a stable population to build up, but also the new species must be able to fit into a niche not already occupied by an endemic species or compete sufficiently to establish its own niche in the presence of an endemic competitor. No sparrow of the genus I have been studying was already present on the Azores, Madeira or the Canaries before the arrival of the invaders, though a sparrow of a related genus, the Rock Sparrow, was present on both Madeira and the Canaries. (The twenty species of sparrows that I am studying belong to the genus *Passer*. The rock sparrows, of which there are six species, are placed in *Petronia*, rather more distantly related, but still in the same family.) The Rock Sparrow, as its name implies, is a bird of much wilder places, usually shunning inhabited parts although it does penetrate cultivated areas. In the absence of competitors it did, however, occupy the more built-up area on both Madeira and the Canaries, though following the invasion of the Willow Sparrow it has now retreated into its more normal habitat.

My interest in visiting these island groups was to see exactly how the Willow Sparrow fitted in. In most of its range the Willow Sparrow is sympatric (that is it has an overlapping range) with the House Sparrow, but, as in Malta and Sardinia, my visits provided another opportunity of seeing how the Willow Sparrow behaved in the absence of its close relative. Once an isolated population becomes established on islands, where the ecological conditions may be very different from those where it originally came from and competition from species with similar life styles is reduced because of the smaller variety of species found on islands, the birds become subjected to different evolutionary pressures and there is the possibility of differentiation from the ancestral stock.

The Willow Sparrow is now well established and widespread in the Canaries. On Tenerife, which I visited in December 1971, it was present not only in the lower-lying cultivated land, but was also common in the towns and villages. On Madeira, on the other hand, I found the species in November 1976 to be confined to a coastal strip from Ribiera Brava, 15 kilometres to the west of the capital Funchal, to Punta dos Reis Magos, 8 kilometres to the east of the capital, and occupying the villages and cultivated land. In the town of Funchal it was making much less use of buildings and similar sites than the birds on Malta and Sardinia, perhaps an indication that it has not yet fully adapted to this way of life. It is perhaps not surprising that the Willow Sparrow is absent from the mountainous north coast and the more inland settlements of Madeira, but I should have expected it to have spread along the south to the extreme west and east of the island. Perhaps this absence is a reflection of the relatively short time the species has been there and the effect of competition with the Rock Sparrow. On the Canaries, where the Willow Sparrow has been present for much longer, the Rock Sparrow is now to be found only in the more arid parts not frequented by

Willow Sparrows. On both groups of islands a very noticeable feature was the way that the Willow Sparrow makes use of trees in the town squares for roosting at night, as has already been remarked for Sardinia.

It seems a bit superfluous for me to say much about Madeira and the Canaries as these are such well known tourist spots. The same cannot be said about the fourth Atlantic archipelago, the Cape Verde Islands, which are much less visited. Here, however, I shall concentrate on the two sparrows already under discussion in this chapter, leaving the Iago Sparrow and my comments on the place until later: the Cape Verdes are a very different country and the bird a very special one as far as I am concerned. Both merit separate treatment.

Even less is known of the history of the Willow Sparrow on the Cape Verdes than on the Canaries. Charles Darwin found it on São Tiago, the principal island in the group, when he visited in 1832. However, as there are no records anywhere else of anyone introducing Willow Sparrows, I think they may well have reached the Cape Verdes naturally, even if these islands lie some 2000 kilometres from the present wintering range of the bird in North Africa, though of course climatic conditions at the time of the bird's arrival, when the arid zone of the Sahara was not as extensive, may have meant that the distance was not quite as great as this. The coincidence of date with the first record for the Canaries may even mean the two colonizations can be traced back to the same climatic event.

Whatever the past history, the present distribution is somewhat patchy. On one island, Fogo, where it is the sole sparrow species, it is widespread, nesting in such diverse places as houses in the principal town, São Filipe, to isolated buildings out in the open country and holes in the cliff walls of the caldera that houses the still smouldering volcano at 1500 metres. In contrast, on São Tiago, where the Iago Sparrow is also present, it is restricted to the towns and the more fertile cultivated land. However, on São Vicente I was able to find only hybrids with House Sparrows, though earlier visitors had reported the Willow Sparrow as common, and on Brava it even died out in the droughts of the 1940s, though it has apparently recolonized this island in the last few years.

The House Sparrow is only found on São Vicente. The details of its arrival are not known, but birds were first recorded in 1924 and it seems most likely that they travelled as involuntary passengers on a ship. Mindelo on São Vicente is the principal port on the islands and was at that time much used by shipping from both Europe and North America; thus it is the most likely spot for such ship-borne invaders. What to me was interesting was that despite their having been present for some sixty years at the time I visited, they were still confined to the town and the immediately adjacent farmland and were by

It seems most likely they travelled as involuntary passengers on a ship.

no means common. This virtually desert island and the competition of the well-adapted endemic Iago Sparrow that swarms there are evidently not much to their liking.

One thing that is obvious from my observations on these groups of islands, is the way the Willow Sparrow tends to move into the urban environment in the absence of the House Sparrow, though it by no means exploits this to the extent that the House Sparrow does, or for that matter the Tree Sparrow in the East where House Sparrows are lacking. It is clear that the drive to become a 'house' sparrow is latent in more than one sparrow species. Here we have a third sparrow willing to move into the built-up environment when the others are absent.

His interest was not in the birds, but more in the attractive figures of the nubile girls . . . disporting themselves in their bikinis on the beach.

6: A difficult journey: The Sahara

What went ye out into the wilderness to see?

St Matthew xi,3

One attraction of North Africa to the sparrow buff, the zone of hybridization between House and Willow Sparrows, has already been mentioned in Chapter 3. To the south in the Sahara lies another attraction, the Desert Sparrow. This is one of those perplexing species with a widely disjunct range. It was first described in 1823 from the Nile in northern Sudan about 65 kilometres south of the Egyptian border and later found to extend westwards from there across the Sahara. Almost 75 years after this another population was reported from Russian Turkestan and subsequently a third was discovered on the western edge of the Great Sand Desert or Dasht-e-Lut in southeastern Iran.

It seems probable that the Desert Sparrow once had a continuous range from North Africa to southern Kazakhstan and that it has subsequently disappeared from much of this area, leaving the three populations described above. This retraction of range is apparently still going on and when I published my book on the sparrows in 1988 I commented that it had become extinct in the Sudan and had disappeared from the eastern Sahara so that it was now to be found no further east than Ennedi in Chad, some 800 kilometres to the west of where it was originally discovered. This elicited some correspondence and I have now details of no less than 10 records for the Sudan west of Khartoum over the period 1959–1990. This is an area of increasing aridity and there seems little doubt that, although my statement that it had become extinct in the Sudan was wrong, there has been a reduction in numbers in this area.

I visited the northern edge of the Rub'al-Khali, the Empty Quarter of the Arabian Desert, in 1982 and it is not surprising that this region is lacking in Desert Sparrows as, with the exception of the Brown-necked Raven, it seemed to be devoid of birds. However, even if this part is now too arid for the Desert Sparrow, that would not seem to be the case in the Sinai Desert and the less arid Negev in southern Israel that I searched in 1980. One would have thought that there would be a transitional region between the completely arid Empty Quarter and the lush Mediterranean coast where the bird could have held out. Whatever the reason, it is not only missing from the Middle East, but has probably also recently become extinct in Iran – the Dasht-e-Lut appeared a pretty unpromising region when I passed along its northern edge on the way back from Afghanistan in 1976. In contrast, the Desert Sparrow still seems to be doing reasonably well in Russian Turkestan.

I have already mentioned the problems I had encountered in trying to get into Russian Turkestan. Iran and the Sudan appeared to be out. So it had to be the western Sahara for my search. The Desert Sparrow had not been reported from the edge of the Sahara in southern Morocco at this time, but in view of its accessibility this seemed to be a good place to sample something of desert travel and there was always the hope that the bird might be there. It would also provide the opportunity of seeing how House and Willow Sparrows are adapted to desert conditions. In November 1969 a group of us headed from Casablanca to Fez and then south towards Ksar-Es-Souk, at the beginning of the desert. We camped in the evening some distance north of Ksar-Es-Souk and, having run out of food, repaired to the town next morning to get some breakfast. What we had not bargained for was that that morning had been declared as the start of Ramadan and all the food shops were firmly shut. The only place we could find open was a French-run establishment and the best we could do there was beer and yoghurt – not the best combination for a healthy, open-air appetite early in the morning. From Ksar-Es-Souk we proceeded south through Meski, Erfoud and Merzouga to Taouz on the border with Algeria. I failed to find any Desert Sparrows in this area, though in 1988 a small breeding colony was found near Merzouga; whether I overlooked it there or this is part of the extension of range to the west I am unable to guess.

The old fort at Taouz was pure Beau Geste; it was easy to imagine members of the French Foreign Legion straining their eyes looking over the shimmering sand for hostile tribesmen. Taouz was also remarkable, though less attractively so, for its flies; I have never encountered so many anywhere else. I am no entomologist and cannot explain why they were so numerous there, and not particularly obvious at other places in the desert, nor why, if this was a regular phenomenon, there were not more of the appropriate birds feeding on this plentiful supply of food. The only way of getting a little respite during the heat of the day when the flies were most active was, despite the discomfort, to shut oneself in one's tent. It was then difficult to decide which was worse: the flies or the oven inside the tent! In contrast the people were far from hostile. In this arid region we had our first experience of the traditional hospitality of the desert. Wherever we went we were invited in for mint tea and almonds, and on one occasion for kebabs (lamb, goat, camel?) and a pipe of hashish as well. As we entered the houses the women silently disappeared and we were served by our host or his eldest son. But, although the native womenfolk took second place, I was surprised to find that with our mixed party the women were always served first, the innate politeness of our hosts recognizing and respecting Western custom. These desert people had a most useful custom in their serving of tea. The first glass was of plain tea without milk, but lots of sugar. Mint was then put into the teapot so that the second glass was mint-flavoured – and most refreshing it was too. After that there was a pause for conversation before the third glass was poured out. This is a recognized signal that hospitality is over. You drink it quickly and make your farewells. How often have I wished that we had such a well-understood convention at home to spur the departure of a tardy guest. I am continually struck when travelling of how little one knows of the culture in foreign parts. As another example, when and from where did tea make its first appearance in the Sahara?

From the desert with an interesting selection of desert birds – larks, wheatears, sandgrouse, but no Desert Sparrows – we moved to Marrakesh where I was able to sample a Moroccan speciality that had long attracted me – sweet pigeon pie. This is a flat pie with savoury pigeon inside, covered with flaky pastry that is coated with almonds and sugar. I was not disappointed. It is a most delicious dish. From Marrakesh we went south again to the desert, through the fertile Sous Valley, where the vegetables, no doubt maturing very rapidly in the warmth and light soil, were quite the most delectable I have ever tasted, and on to Goulimine.

Goulimine is known for its great camel market, and inevitably anywhere with a collection of camels is rather a sordid place, so it was pleasant to get out again into the cleanliness of the desert. Our objective was Aereora, a deserted fort on the coast at La Plage Blanche where the Saharan sands finally disappear under the water of the Atlantic. Although the fort was no longer in use, a Moroccan caretaker lived there with his family and once again we were hospitably received with tea and an enormous pile of roasted almonds on a brass tray. It was from the wall of the fort that I was watching shorebirds through my telescope, when our Moroccan host asked, '*Shifti bint?*' His interest was not in the birds, but more in the attractive figures of the nubile girls of our party disporting themselves in their bikinis on the beach. The girls provided an endless source of attraction for the native males. Later a passer-by on a camel asked if we would exchange one of them for four camels. Not having much use for camels nor any idea of their market value, and to the relief of our companion, who did not much fancy the malodorous nomad, we rejected the offer.

I was disappointed, but hardly surprised, that I did not find any Desert Sparrows, although the House Sparrow was common at all the desert oases we visited, congregating round the area of the gardens, showing how well it is adapted to desert conditions. At Aereora I was interested to find a flock of twelve Willow Sparrows in a wadi with a few scrubby acacias, a very different habitat from the moist areas I normally associate them with; they must have been winter migrants, as we were well outside the breeding range of the species, and at an easy stepping off place to Lanzarote in the Canaries, less than 250 kilometres away. In Goulimine, which lies at the southern extremity of the breeding range of both these sparrows, the birds were mostly hybrids.

This was my first experience of genuine desert and I was quite captivated by it; so much so that since then I have gone out of my way to visit as many of the world's deserts as possible. It is difficult to explain this fascination. To a naturalist there is the excitement of going to an area that has been so little affected by man and seeing something of the way the animals are adapted to existence in the harsh environment. Getting down to the real basics for survival – water and food, and one's ability to overcome the difficulties that always arise is something else. Lying out at night looking at the extraordinarily bright stars and being very conscious of the complete absence of noise – a real luxury in our modern world where it is very difficult to escape the pollution of man-made sounds – also has its attractions. But there is something else that defies explanation and has to be experienced first hand. It is not surprising to me that so many of the world's great religions have seen their origins in desert regions.

Having got hooked on the desert and being more determined than ever to see the Desert Sparrow, I made an extensive journey in May 1971 through the Algerian Sahara. We flew to Algiers to pick up an internal flight to the desert town of Ghardaia where we were to meet our party of three Land Rovers. Unfortunately, as the result of a sandstorm further south, there was a delay and we had to spend some hours in the transit lounge at Algiers airport. The number of transit passengers greatly exceeded the available seats, but we were not allowed to move to the unoccupied international lounge, where there was more room, nor, although we were well provided with sleeping bags, was it acceptable to the authorities that we should stretch out on the floor, though why this was so we could not find out from the armed guards who prodded us with their weapons when we tried to lie down, so we had to spend the long night on our feet. The dust had settled by the morning and on arriving at Ghardaia after a short flight the stupidity and discomforts of the night were quickly forgotten as the desert had me once again under its spell.

There are three traditional north–south routes across the desert: from Bechar to Gao in Mali, from Laghouat to Kano in Nigeria and from Biskra to N'Djamene in Chad. Our plan was to follow the centre route to El Golea, then to divert across to join the western route at Timimoun and follow this route south to Adrar before striking back to rejoin the centre route at In Salah and continuing on to Tamanrasset. Each of these oasis towns has its own special character, partly through their settings, partly through differences in architecture. I found Timimoun particularly attractive. The region to the south of Timimoun, known as the Valley of the Palms, used to be a very fertile one, being well irrigated by the underground canals known as *foggaras* that brought sweet water from a natural collecting point at the base of some distant hills. These underground tunnels, which run for miles with a gradual slope to provide the drainage, are an incredible feat of engineering. They can be recognized from above by the regular heaps of spoil that march in straight lines across the desert out of sight. In the loose ground they require regular maintenance to prevent blockage and the unpopular and dangerous task of digging and keeping them open used to be carried out by slaves. With the disappearance of this forced labour the *foggaras* are gradually silting up and the Valley of the Palms is drying out and returning to the desert. One *foggara* that I saw was still flowing freely at a rate of several litres per minute and where it came close to the surface one could dip one's hand into water that was surprisingly cold in the hot surrounding desert.

Up to Tamanrasset we had followed a well-established, though in places somewhat ill-defined, piste, but our next west-to-east traverse to Djanet through the Ahaggar Mountains was well off the beaten track. The Ahaggar is a fascinating area that even a seized differential on one of our three vehicles could not dispel. Although we were carrying a spare differential on board, the difficulties of field maintenance, to say nothing of a violent sandstorm that suddenly blew up, inhibited our attempts at replacement and we had to struggle forward with only front-wheel drive and the propeller shaft disconnected. Soft sand was the major problem, but with manpower and the help of

The author at the magical oasis of Djanet, southern Algeria.

the other vehicles when that was not enough we eventually made Djanet. It brought home to us the sense of the desert regulation that one must always travel in convoys.

This tribulation no doubt added to the enchantment of Djanet as an oasis, but I think that in any circumstances with its palms and greenery it would be a magical place to the traveller. Djanet has another attraction, the Tassili-n-Aijer, a rocky plateau lying to the north. The Sahara is not exactly crowded with vehicles – in the two days from Tamanrasset to Djanet we had not seen another one – but the Tassili-n-Aijer, perched 700 metres above and surrounded by rocky walls (Tassili means 'a rocky plateau' in Berber), has the added bonus that it is inaccessible to anything on wheels. So we left our vehicles, one to have a new differential fitted, and departed on foot with our baggage on mules to explore something of the plateau. Much of it is a fascinating, sand-eroded moonscape, now virtually uninhabited, though at one time an important area, judging by the rock paintings at Tin Tazarift which show vivid hunting scenes and many recognizable animals, such as elephant, rhinoceros and giraffe, now only to be found much further south.

The rock paintings, which are on overhangs not in caves, are in fact incisions in the rock that have been filled in with ochre, rather than flat paintings; they are still not satisfactorily explained. Presumably the fact that

Sand-eroded moonscape, Tassili Plateau, southern Algeria.

they are of animals no longer present in the area means that they date back to the pluvial period that occurred about 10,000 years BP (before present), before the desiccation to the present desert that began about 5000 years ago. But in that case, why have they persisted so long in an area where there has been so much erosion by wind-blown sand? Another interesting occurrence on the plateau is a cypress, *Cupressus dupressiani,* that is now reduced to a small number of specimens and is found nowhere else. These trees are individually numbered and protected, but at best their survival can only be very limited as there is no longer any natural regeneration.

A paradox of the vast, sparsely inhabited region of the Sahara is its parochial nature. The camel is the workhorse of this area, but although domesticated it still has a will of its own. I am told that it is only possible to work a camel for six months; if this is exceeded it gives up and dies, apparently a mental decision as veterinarily nothing can be found wrong with it. For this reason one regularly comes across groups of unattended camels browsing across the desert. Once they have been mentally (or physically?) restored they have to be rounded up and the owner sets out on foot to look for them. During our enforced stop due to the breakdown in the Ahaggar, one such owner approached and asked us if we had seen a group of eight camels including a distinctive white one. He did not seem unduly surprised when we

Once they have been mentally restored they have to be rounded up and the owner sets out on foot to look for them.

told him we had and walked on content with our directions of how to find them.

Djanet was our furthest point south, and from there with three serviceable vehicles we once again headed north, again without sight of my prey, though I have since been told that it is numerous between Djanet and Ghat in Libya. This time our route lay over the Grand Erg Oriental with its mountainous dunes of sand, through the oilfield region of Hassi Messaoud, to the oasis town of Ouargla and finally to Ghardaia our pick up point on the way home, still on the look-out for the elusive Desert Sparrow. Hassi Messaoud is not a traditional oasis settlement, but a modern town created by pumping up water from Savarin's Sea, a body of water that occurs in a permeable stratum above the petroleum. In normal circumstances the cost of drilling for the water would not justify its extraction, but Hassi Messaoud with its gardens shows how potentially fertile the area is if plenty of water can be provided.

Mention of the Grand Erg Oriental brings to mind two facts about the desert that are not well understood. Most people imagine the Sahara to be a vast area of unrelenting sand. In fact, the two great sandy areas of the Grand Ergs Oriental and Occidental, with their dunes reaching up to 200 metres high, make up less than 20% of the surface of the Sahara, which otherwise has a varied scenery, ranging from flat pebble reg, on which one can drive for miles at up to 45 m.p.h. in any direction with one's hands off the wheel, to hamada that is largely flat undulating exposed rock, and mountainous regions such as the Ahaggar that rise to over 3000 metres. All of this is interspersed with wadis lined with acacias and bushes, such as *Calotropis* with its purple flowers and large green seed cases about four times the size of a ping-pong ball and about half the weight. There is a surprising amount of vegetation and with it insects, reptiles and mammals as well as an interesting selection of birds.

The desert is still, however, a harsh place and it is necessary to ask the authorities for permission to travel and to inform them of your intended route so that in the event of non-arrival at your destination the necessary rescue operation can be mounted. In addition, one must take the simple precautions of having enough water, food and fuel to survive the journey. Nevertheless, the Sahara has claimed many lives. Perhaps the most surprising fact about it is the answer to the question as to what has been responsible for the greatest number of deaths of Westerners since they began to penetrate it at the beginning of the 19th century. Likely candidates would seem to be hostile natives, dehydration or disease; but no: the unlikely answer is drowning. Desert is conventionally defined as having a rainfall of less than 250 mm of rain per year; this means that the Sahara begins south of the Atlas in the north with a rather ill-defined boundary 100 km wide and reaches in the south to the Sahel in the west and the Sudan in the east, extending from the Atlantic right across to the Red Sea. In the centre the rainfall is, of course, much lower and is so irregular that to talk of an average annual figure gives a somewhat misleading impression, disguising the fact that in some places it may only rain once in 10 years and such rains as do fall may do so in violent storms; for example, in one storm recorded at Tamanrasset 36 mm of rain fell in only 40 minutes. In such a storm the rain runs straight off the baked ground and flows as a mighty wave along the wadis carrying everything before it. Some friends of mine, camping one night in a wadi, actually lost one vehicle, and nearly their lives, as the result of a storm some 25 kilometres away; they heard the row of the advancing water and had just time to scramble out of the wadi before everything was swept away. A golden rule of desert travel is not to sleep in wadis, even if they do look attractive and provide a shelter from the wind-blown sand. My only experience of rain in the desert was one night camping near Djanet. I was sleeping in the open and awakened to gently falling rain. It was a most odd experience; the rain was so warm and the atmosphere so dry that the drops evaporated as quickly as they fell and there was no sensation of getting wet. It was the first rain that had fallen there for over five years and obviously was not going to do much for the soil.

Now what about the sparrows? I had been assured that I would have no difficulty in finding Desert Sparrows along the route we took, but it was not in fact till our last morning at Ouargla, after three weeks and 5000 kilometres, that on my last dawn walk I saw my first pair. Delightful birds, the male with its small black bib obviously a sparrow, but with otherwise silvery grey plumage, and the pale sandy female, they were both clearly adapted to the desert habitat. I was disappointed not to have seen something more of their behaviour, but delighted that the journey was at least not in vain as far as the Desert Sparrow was concerned. I have since spoken to other birdwatching travellers in this area and they also had difficulty in finding Desert Sparrows. Perhaps this is the one sparrow that is really losing out.

However, there were other sparrows to be seen. Willow Sparrows were at the palmeries both in Timimoun and Djanet – a flock of fifty or so at the latter place that I had to stalk for about an hour before I was able to confirm the identification. Both these places are well to the south of its known breeding range, adding confirmation to its great capacity for wandering and lending support to the possibility that it spread unaided to the Canary Islands. House Sparrows were seen at El Golea, Adrar, In Salah (in both the village and the palmery), Ouargla and Ghardaia, though somewhat surprisingly not at the modern town of Hassi Messaoud that has grown up at the oilfield. These records are to the south of the hybrid zone I have mentioned in Chapter 3, though Wilhelm Meise had described hybrids from Ouargla and Ghardaia in 1936. Are the House Sparrows taking over? Have they eliminated any trace of the Willow Sparrow from that area?

My wife meantime was mentally checking up the insurance to see what she would collect as a widow.

7: Going for gold: Senegal

To gather gold
At the world's end I am sent.
Rudyard Kipling, '*The Tunes on Weyland's Sword*', 1906

The Golden Sparrow occurs in Africa in a narrow latitudinal band, at its widest less than 8°, lying just to the south of the Sahara, stretching from the Senegal Sahel in the west to the Red Sea coast of Sudan and across the Red Sea into Yemen and the extreme southwest of Saudi Arabia. Senegal is the most accessible of the countries in which it occurs and, moreover, the bird has been studied extensively there by ornithologists from the French overseas aid organization, ORSTOM (*Office de la Recherche Scientific et Technique Outre-Mer*), because of its possible activity as a pest of agriculture.

I got in touch with Dr Gérard Morel, who was in charge of the Station d'Ecologie de l'ORSTOM at Richard Toll on the Senegal River in the north of the country and who with his wife, Marie-Yvonne, had carried out much of the work on the bird. We made arrangements to go to Senegal in September 1984, when we expected the birds to be at the beginning of their breeding activities. Gérard phoned just before we left England to say that he had some Golden Sparrows in the grounds of the research station and everything seemed to be falling in line with the plan. I said I would call him from Dakar when we arrived to get our final instructions as to where it would be best for us to go.

Dakar is not my wife's favourite airport. We had spent a night in Dakar the previous year on our way to the Cape Verdes, when the importunity of the would-be porters and the taxi touts ended in a scuffle before we succeeded in getting a lift to our hotel. This time we decided to stay the first night in the airport hotel and pick up a hire car the following morning when we would be setting off north. We changed some money, booked into the hotel and returned to the airport bar to have a drink. There we were joined by a well-dressed youth who suggested we should give him some money to help him feed his large family, saying he knew we had plenty as he had seen us changing traveller's cheques at the bank. Fortunately the barman recognized him and he was quickly escorted from the scene. That apart, we found the local people very friendly, though every time we stopped the car a gang descended upon us offering to clean it. That did not get very far with us, as being a hire car we saw little benefit in paying to have it washed several times a day. We refused to be intimidated by the ones who started to wash it before asking if we were interested and this was accepted cheerfully.

I called Gérard right away from Dakar to learn that the Golden Sparrows had disappeared. But, not to worry; if I came straight up we should have no difficulty in locating them. The Golden Sparrow is highly social with roosts containing up to a million or more birds and breeding colonies frequently consisting of more than 50,000 pairs. The birds in these immense gatherings are noisy, so no problem was envisaged in finding them. At Richard Toll, Gérard took us on an extensive tour round, finishing at dusk at a sugar-cane plantation where the birds had been roosting in large numbers. Not a sign of a single Golden Sparrow did we see. It was back to the drawing board. The bird is nomadic, settling down to breed in localities where the conditions are suitable. Gérard gave us a specification for a breeding colony: scattered acacia savanna with emergent green vegetation below the trees. The Morels were returning to France in a few days for their annual leave and all we had to do was to drive round northern Senegal to look for suitable conditions.

We started by going east along the Senegal River, but it soon became clear that everything was much too dry there. There were plenty of acacias, but nothing but bare sand between them. We then moved south where conditions were somewhat better, though in fact it took us nine days before we found our first breeding colony; after that we found five more and Marie-Yvonne put us on to a sixth that she spotted when driving from Richard Toll to Dakar airport on her way back to France. Most of the colonies were further south than had been previously reported, reaching almost to the limit of the thorn tree savanna zone, a reflection no doubt of the continuing Sahel drought pushing the growth of the ground vegetation that creates favourable conditions for the insects on which the Golden Sparrows depend for the rearing of their young, further to the south.

The Golden Sparrow is a striking bird. It is one of the smallest sparrows, weighing only 14 grams, but the male makes up for this with its bright canary-yellow head and underparts, chestnut mantle and back. The formation of a breeding colony is a striking spectacle with the brilliant males disputing for their nest sites and chasing each other through the green acacias against a noisy background of excited *chitta chitta churr chitta chit* calling in the otherwise quiet savanna.

In searching for the sparrows we covered quite a bit of Senegal, mostly in the semi-arid northwest, where we could see the effects of the prolonged dry phase that was increasing the aridity of the Sahel and making life very difficult for the indigenous peoples. The furthest east we penetrated was to Linguère, where the tarmac road ends and the thorn savanna really takes over. Here we asked a policeman if there was anywhere we could stay. He was an enormous man, but he managed to squeeze into our small car and took us to what he described as an excellent hotel – the Centre du Tourisme de Linguère. At first

Hotel du Centre du Tourisme de Linguère, Senegal. The proprietor and the author.

sight it looked little more than a bar with enamel-topped tables and a few rooms at the back, but there was no alternative and nowhere else within range. We booked a room and went to explore east along the dirt road looking for birds. We returned at dusk and were asked what time we required dinner. Eight o'clock was agreed and we returned to our room to tidy up as best as we could with a jug of water. The shower in the bathroom attached to our room showed no signs of having seen any water for quite some time, another forcible reminder of the desperate water situation.

At eight we prepared to go to the brightly lit bar with its striplights and enamel-topped tables; but no, we were led out to an enclosed courtyard where there was a beautifully set table with gleaming cutlery on a damask cloth illuminated by a spotlight. We had a delicious meal under the stars, cooked by a lady called in specially for the occasion, watching the distant flashes of summer lightning and the geckos feeding on the flying insects that rashly approached the spotlight too closely and ended up on the ground below. The French have certainly left their impression even in the depths of the African bush. European visitors are nowadays few and far between. Having found the Golden Sparrows and watched their breeding activities, my thoughts turned to the Morels back at home in France, perhaps at this very time drinking their special Golden Sparrow wine at their house in Normandy.

Gérard Morel's Golden Sparrow wine.

This pleasant memory was rather spoiled by an incident on our way back. We were driving through a small village, slowly because we were not sure of our way, when an armed policeman jumped out from behind a tree and we were fined for exceeding the speed limit. There was little use in arguing and we made our small contribution to Senegalese funds; at least we supposed so as we got a receipt.

Journeys in Africa can be a little different from those in Europe. In 1972 I was looking for Swainson's Sparrow in Ethiopia. Despite warnings against bands of *shufti* that were said to be operating in the north of the country, particularly in the Province of Eritrea, Ethiopia was an extremely pleasant place to visit in 1972, with very friendly people and magnificent scenery. In Eritrea, where the bandits were most active, we were not allowed to camp and in the rest of the country it was recommended that we camped in villages for protection, but there were such numbers of children in the villages that we thought the *shufti* the lesser hazard and preferred the open country.

In the villages the people appeared happy and very unsophisticated, to such an extent that they crowded the roads and completly ignored the horn. The only thing to do was to drive very slowly forward, but even at the slowest pace it was impossible not to push the odd person unceremoniously into the ditch. This caused so much hilarity that the bystanders fell about the road laughing. sn09Travel was also exciting even without the risk of meeting up with bandits. The gorge of the Blue Nile north of Addis Ababa is very steep and in places the road is built out from the valley wall on a viaduct that overhangs the river below. At one of these there was the sign: 'Caution! Viaduct ahead has

Viaduct over Blue Nile Gorge, Ethiopia. Drive fast would seem to be a safer option.

At Khartoum Airport a pair of House Sparrows caught moths attracted to the lights.

started moving. Drive very carefully and slowly.' It was not clear what precautions one could take apart from ignoring the warning and getting across as quickly as possible before the road fell away. This we did.

A surprise awaited me when I arrived in Senegal. The first bird I saw was a House Sparrow. I was unaware that it was anywhere on the west coast of tropical Africa. Evidently they had turned up unnoticed in Dakar. The exact date is not known; the first definite proof of their presence was in 1978, though there is circumstantial evidence that the first arrivals date back to 1971 at least. Since then they have spread rapidly through the Europeanized towns of western Senegal, north along the Senegal River, where they are now well to the east of Richard Toll, and south to Banjul, the capital of The Gambia. It seems likely that the colonization of Senegal began with some ship-borne immigrants, I suspect from South Africa as the Senegal birds closely resemble the House Sparrows introduced there at about the turn of the century. I was delighted to find that Dr Marie-Yvonne Morel was closely monitoring their spread and breeding ecology in Senegal, so that we should get an understanding of how this bird of essentially temperate and subtropical latitudes adapts to a tropical environment. (Unfortunately, the Morels have had to return to France since this was written, so Marie-Yvonne's study has come to an abrupt end.) Once again I was impressed by the bird's obvious success. They positively swarmed at Dakar Airport, flying in and out of the terminal buildings where they had their nests. The presence of sparrows always helps me to while away the long hours at airports. Once at Khartoum Airport I was able to watch a pair of House Sparrows feeding their young; the remarkable thing about this was that it was at two o'clock in the morning, the parent birds catching the moths attracted to the lights and taking them to their young in the nest. It is certainly an adaptable bird.

There were no House Sparrows in Banjul, or Bathurst as it was still called, when we were there, in 1973. The Gambia in January was a welcome change from the cold of England. It appears to have been developed as a tourist area by the Swedes – why was this opportunity not taken by the British in their former colony? Many of the visitors too appeared to be Scandinavian and when we made a trip up the river, staying the night at Tendaba, we were the only British guests. The gentleman who made a speech of welcome (in English) apologized that he could not speak Swedish and was evidently unaware that there were any British people in his audience. I went up to introduce myself afterwards and he turned out to be the BBC correspondent in The Gambia. He was with the local MP and after dinner we received an invitation to join the party, which largely consisted of the MP's wives (?). A hundred kilometres up this tropical West African river we were treated to French Champagne – a bit different to the reception given to my fellow Scot, Mungo Park, on his exploration of West Africa almost two centuries earlier, but a somewhat dubious item on which to be spending the country's precious foreign exchange.

The Gambia presents marvellous opportunities to the bird photographer. Nowhere on my travels have I found birds to be so numerous, tame and approachable. At that time I was using a 400 mm telephoto lens that would only focus down to 6 metres. The Gambia was one of the few places where I found that I had regularly to be fitting extension tubes to be able to photograph birds at closer range. I overcame this temptation with a crocodile, which I stalked up to the focusing limit. Even at 6 metres I wondered whose nerve would go first. Neither did and I got a close up of a nice mouthful of teeth that would look well on the wall of a dentist's waiting room. My wife meantime was mentally checking up the travel insurance to see what she would collect as a widow.

It did not take me long to locate the Somali sparrows . . . on a tree behind the nunnery.

8: Paradise: Northern Kenya

Get thee to a nunnery.
Shakespeare, *Hamlet,* III, i, 124

The Somali Sparrow as its name implies occurs in Somalia; the range extends to northern Kenya and the extreme south of Ethiopia; of these, Kenya seemed the best place for a visit. All the same it is not a common bird in Kenya – at the time I was planning my visit there was not even a single breeding record – and what records that were available suggested the area of Lake Turkana as the most likely for a search. Lake Turkana is not the most accessible place so I persuaded a number of birdwatching friends to join me in an expedition: not that a visit to Kenya requires much persuasion; it is a marvellous place for the naturalist interested in birds and mammals. Nine of us set off from Heathrow in February 1982. We had organized a four-wheel drive Bedford truck to meet us in Nairobi, complete with camping equipment, driver and cook, so that shortly after our arrival we were on our way to Lake Nakuru where we were to spend the first night. The plan was to work up the Rift Valley, stopping at Lakes Bogoria and Baringo on the way to Lake Turkana. One of our party had attempted this journey the previous year on his way to the Sudan, but the trip had come to an abrupt end at Lodwar when he contracted cholera, despite having had the appropriate vaccination against it before departure. He was lucky that there is a small mission hospital at Lodwar, where he was quickly admitted and put on a life-saving drip. It was a useful reminder to us to take no chances with unboiled or untreated water. The previous time I had been in Kenya with Derrick Spinks he had mistakingly taken a drink of paraffin from a jerrycan he thought to contain water and lived to tell the tale not much the worse. Derrick was obviously a survivor and a good man to have with you on a trip.

It is a joy on these expeditions to have a cook in the party. It means that one can get up at dawn and have a few hours with the birds when they are at their most active and then return to a large breakfast ready and waiting. Thomas, the cook, had seen it all; he was very lame from an encounter with a buffalo and on another occasion had wakened up to see a lion in his tent. He had rather a predilection for the local hooch so that we soon learned not to pitch camp near inhabited places if we wanted a good breakfast the following morning. That helped to keep the cooking on an even keel, though it was remarkable that, however far out in the bush we were, and despite the fact that after his animal encounters he was no great looker, Thomas always seemed to manage to find a 'sister' to keep him company for the night. Thomas's cooking abilities over a wood fire – and those also of Chegi whom we had had on a previous Kenyan trip – were, to say the least, remarkable. My wife, who is a keen baker of bread using a precisely temperature controlled electric oven, was fascinated to see bread baked on an open wood fire. Chegi even managed to bake sponges. He was a bit of a martinet and you quickly fell out of his favour if you did not eat up all you were given. As we were hungry living in the open air and the food was excellent, this was not much of a problem.

In marked contrast to my previous visits to Nakuru, there were few flamingos to be seen round the lake. In 1978 on our approach to Nairobi in the plane a clear ring of pink from the immense numbers of flamingos could be seen round the lake from our height of about 7000 metres. In 1982 they had moved to Lake Bogoria. This is one of the ornithological sights of the world. It is difficult to estimate numbers on a fleeting visit, but the consensus of our party was that there were more than 100,000 birds present on Lake Bogoria. Apart from the flamingos, Lake Bogoria is an attractive place, cradled in the mountains, with boiling-hot springs of clear water near our camp site, and remote from habitation. Even Thomas failed to find a 'sister' there.

Lake Baringo is quite a contrast, with an airstrip and a large hotel complex on the west shore of the lake and plenty of facilities for tourists, even to the extent of having a resident ornithologist to take visitors out on birdwatching expeditions. Terry Stevenson had been there for four years and I was delighted to find someone with up-to-date knowledge of the avifauna of the

A tent is not too much of an obstacle to a two-ton hippo in full flight.

area. I could not wait to question him about the Somali Sparrows. He replied that although he had travelled extensively over north Kenya in the four years he had lived there, he had not seen a single Somali Sparrow – and here there was a long pause – until three weeks ago when he found a flock of about thirty birds in Kapedo, a small village about 50 kilometres to the north. This was good news and rather unexpected as it was about 150 kilometres south of the previously reported range in Kenya. Terry was also able to tell me that the first definite proof of breeding of the Somali Sparrow in Kenya had been obtained the previous year. This was in the Kaisut Desert, 4 kilometres north of Korr to the southeast of Lake Turkana.

We camped on a flat piece of ground beside the lake only to find that this was on a hippopotamus route when they came out of the water at night to feed. They do not worry about the tents; the important thing is not to frighten them when they are out of the water as their immediate reaction is to make a bee-line back – a tent is not much of an obstacle to a two-ton hippo in full flight! It is more comfortable to watch them floating in the water during the day, though even then it does not do to approach them too closely.

A walk with Terry at dawn gave us a good look at the local scene and produced a number of new birds, but needless to say we were quickly off after breakfast on our way to Kapedo. This was a small village on the Suguta River and in arid semi-desert country. It did not take me long to locate the

A moment of triumph. First sighting of Somali Sparrow, Kapedo, Kenya.

Somali Sparrows: the thirty birds seen earlier by Terry Stevenson were still there, in a tree behind the nunnery. This was the highlight of the trip for me and it was easy to observe the birds, which kept close together as a social flock, perching on the trees and houses and feeding on the ground, even going inside tents, belonging to a detachment of soldiers, that had the flaps open. It was too early for breeding when we were there and unfortunately Terry Stevenson was not able to visit Kapedo during the breeding season, which according to the one breeding record for Kenya is in the rains in May, the same time that it breeds in Somalia. However, judging from the displays I saw and the fact that the males had black bills – the bills of male sparrows change from horn-coloured to black as they come into breeding condition – it looked pretty certain that they were going to breed in Kapedo. Terry did manage to get there in February 1983 and again found a flock of thirty birds, so it looks as though this is a resident population. He also found a smaller flock of around twenty birds about six kilometres away; it would seem that the birds in Kapedo were no fluke and that this area is certainly within the range of the Somali Sparrow in Kenya.

Grévy's zebra: a typical northern Kenyan semi-desert species.

After Kapedo we decided on a change in plan. The original intention had been to go to Lodwar and Ferguson's Gulf on the west of Lake Turkana, but with the breeding record for the previous year in the Kaisut Desert we made up our minds to go to the east of Lake Turkana through the Kaisut to Marsabit. Marsabit lies on a forested mountain that rises between the Kaisut and Dida Galgalla Deserts. We camped in the reserve that lies to the south of the town beside the so-called Lake Paradise. This lake is in the crater of an extinct volcano, surrounded by dense forest. After the desert and its heat the sight of the water and the lush vegetation is a marvellous thrill, but this is mitigated to some extent when one gets to the edge of the lake. The water is thickly covered in green slime; even after boiling for twenty minutes and skimming off the assorted dead fauna it still looked very unappetizing and we restricted its use to washing ourselves and the dishes. The person who named the lake must have had a longer time in the desert than we did and been in worse straits when he arrived. I failed to see any more Somali Sparrows, but it was nice to have our first sighting of the very lovely, and also rather rare, Grévy's zebra.

My wife was happier with the water in the Masai Mara National Park that we went to afterwards. We camped there beside the Talek River, a beautiful clear stream running over a gravelly bottom, and almost before the tents were up she was happily scrubbing away at our dirty clothes. Just in time! The next morning the river was dry and this time there was no one to complain to about the waste pipe!

The Somali Sparrow has been described as the 'house' sparrow of Somalia, living socially in the towns and villages and nesting in holes in buildings, though it also occurs away from habitations and is noted particularly from sea caves on the coast. In appearance it bears a distinct resemblance to the Cinnamon Sparrow of Asia, of which more later, but I was most struck by its behaviour – it was difficult to convince myself that I was not watching a flock of House Sparrows. The House Sparrow has a very characteristic display before and during the breeding season in which a number of males chase and hop around a single female with drooped wings and with head held high showing off the black bib. When I was at Kapedo I saw two exactly similar displays by the Somali Sparrows. I had not previously seen this display in any other species of sparrow, except once in Malta. It will be remembered that the sparrows in Malta are not completely pure, but are at one end of the hybrid zone between the Italian and Willow Sparrows discussed in Chapter 3 and have some House Sparrow blood.

Further evidence of the close relationship between Somali and House Sparrows is shown by the collection of a hybrid between these two species by John Ash near Mogadishu in Somalia in March 1980. This raises the question of where the bird came from as the ranges of the two species do not overlap. John Ash suggests that the House Sparrow parent may have been a casual immigrant aboard a ship. This probably also applies to the three House Sparrows (two cocks and one hen) that turned up in Mogadishu in December later that year. As already mentioned, hybridization with such small numbers is more likely as the birds may well have difficulty in finding a mate of their own species. There have also been a number of instances of House Sparrows turning up as ship-borne passengers in this way, the best recorded example being the colonization of the Falkland Islands in 1919 by a party of House Sparrows that hitched a lift from Montevideo in Uruguay on a fleet of whaling vesssels and disembarked at Port Stanley. So ship-borne arrivals of House Sparrows in Somalia are quite possible.

The close relationship of the Somali Sparrow with these northern sparrows raised all sorts of interesting speculations in my mind as to the origin of the species. I believe that the sparrows originated in Africa. Was the Somali Sparrow, or a close relative of it, the sparrow that first made the journey into the Palaearctic region? Did it have an immediate common ancestor with the House Sparrow? Could the Somali Sparrow have evolved in the north and be a secondary invader to the African homeland of the sparrows? To raise questions of this sort was the real reason for my searches. Northern Kenya had lived well up to its promise.

My wife found it a long day under a shady tree . . . while I watched the birds.

9: A solitary sparrow: South Africa and Botswana

At least, when we get outside the cities
we get out of sparrowdom.

Fraser's Magazine, January 1880

Rufous sparrows occur in four widely scattered populations on the African mainland. In addition, there are two further populations on islands off the coast: one on Socotra and Abd-al-Kuri Islands in the Indian Ocean off the Horn of Africa, the other on the Cape Verde Islands in the Atlantic Ocean off the west coast.

Several times I had had fleeting glimpses of Rufous Sparrows during visits to Kenya, but never the opportunity of making prolonged observations. South Africa looked a better place from the point of view that there are a number of first-class ornithologists there who I felt would be able to give me some precise information on where best to go to see the bird. The plan I finally came up with was to start at the Ornithological Research Station at Barberspan in southwestern Transvaal. The Rufous Sparrow, or Great Sparrow as it is known in South Africa, has been recorded as an infrequent visitor to the Barberspan Reserve, but my main intention in going there was to have a look at another species, the Cape Sparrow, and hopefully pick up some hints from

Dr David Skead of the Transvaal Nature Conservation Division, who lived and worked there, on the best approach to finding Great Sparrows in Botswana, one of the likely places for them. After that we arranged to go to the flood plain of the Nyl in central Transvaal, where I hoped that Warwick Tarboton, another outstanding field ornithologist in the Transvaal Nature Conservation Division, would be able to guide me to the right places. Incidentally, 'Nyl' is the Dutch word for Nile. At one time the Nyl was presumed to be the upper waters of the Nile, the source of which was much sought after by early European explorers. They were a mere 3000 km too far south.

Enquiries through travel agents and bucket shops established that the cheapest way to get to Johannesburg was by means of Air Zaïre. This involved a flight from Heathrow to Brussels to pick up the Air Zaïre flight to Kinshasa where there was a TAP Air Portugal connection to Johannesburg. The kindest thing to say about Air Zaïre is that it got us to Kinshasa more or less on time. It is difficult to find a kind thing to say about Kinshasa Airport. No doubt it was fine when designed and built, but years of neglect had made it rather sordid and no place to arrive in the small hours of the morning with uncertainties about a connecting flight. Airlines and aiport authorities the world over are skilled in passing on the minimum information to the paying passengers. At Kinshasa our passports and tickets were taken from us and we were herded into a transit lounge – like many international transit lounges with all the seats already occupied by procumbent figures apparently having given up hope of moving on. There was a complete absence of officials from whom one might have tried to get information, even to the extent that some of the passengers wandered out on to the tarmac to have a look around without being stopped. I prefer to keep a low profile in such circumstances. On another occasion, on reboarding our plane at Khartoum on our way to Nairobi, an armed soldier at the foot of the steps to the aircraft demanded my wife's passport, which he proceeded to examine carefully page by page before handing it back, apparently satisfied despite the fact that he had been holding it upside down. All in fact that was on offer at Kinshasa was beer, not our first choice of refreshment after a fitful night's sleep on a plane and no breakfast. Dawn brought some relief as there were birds to watch on the tarmac even if there was no other activity there. It got hotter. In that spirit of camaraderie that inspires the British in adversity, we began to identify the hopefuls bound for Johannesburg and talked to each other as we passed backwards and forwards along the veranda of the transit lounge. All this did was to generate rumours: that plane on the other side of the tarmac was the Portuguese one affected by a technical fault – that euphemism of the airlines to cover any delay; there was no connection to Johannesburg – what do you expect with a cut price fare? Then a plane arrived and in due course a girl appeared with a handful of passports and tickets, and called out some names. People detached themselves from the lounge, were escorted across the tarmac and departed for some unknown destination. At last another plane arrived and with my field glasses I was able to decipher the magic TAP on the tail. We could not decide if it was on time or late, because although we knew the scheduled time of arrival in Johannesburg we did not know the local Zaïre time – the clocks in the departure lounge had long since given up – nor the duration of the flight. But we were back in the luxury of a European plane, with friendly, communicative stewardesses and a most welcome, if somewhat belated, breakfast.

As it turned out we were only two hours late at Johannesburg; but with a drive of over 350 kilometres to Barberspan it was obvious that we were not going to arrive until after dark. The Sunday roads round Johannesburg were quiet, but with the warnings of heavy on-the-spot fines if one was caught exceeding the 90 km/h speed limit, we drove circumspectly. In fact, just at dusk I seemed to go through a police trap, as immediately we passed uniformed figures dashed out into the road and flashed lights. Not being sure whether it was the police or an ambush, I thought the best thing to do was just to drive on and hope for the best; this I did and we saw nothing more. As dusk was falling it suddenly got pitch black and we had a magnificent thunderstorm. My wife is not keen on lightning, but even with closed eyes covered by her hands she could still see the lightning flashes. The road was difficult to follow even with the almost continuous illumination of the lightning flashes; it was flowing with water and not surprisingly there was no other traffic about and we began to have doubts about our map reading when we finally saw a sign to the Elgro Hotel, our destination. As a result of a misunderstanding we were not expected, and after a hectic weekend the staff had all gone to bed. However, having made contact we were made most welcome and, refreshed by some beer and sandwiches, we went to bed after some thirty-eight traumatic hours on the way. Sparrow watching can be an arduous activity.

David Skead was away when we arrived. In his absence, John McAllister, the Reserve warden, rallied round and took us under his wing; he showed us over the reserve and pointed out the Cape Sparrows I had come to see, including a pair that were building a nest on the roof of his garage. It was warm and pleasantly relaxing to sit in a comfortable chair in his garden watching the birds through a telescope mounted on a tripod, suitably refreshed from time to time with cooling drinks brought out on a tray, infinitely preferable to what I would have been doing on a bleak November day at home. This was the most luxurious sparrow watching that I have done.

David Skead, when he returned, gave me just the help I wanted. His advice

The late David Skead, field ornithologist, Barberspan, Transvaal, and his wife.

was that if I wanted to be sure of seeing Great Sparrows, Botswana was the place to go to. He suggested that I call on Alex Campbell at the Botswana National Museum in Gaborone, the capital. Little did he realize that growing up as a lowland Scot I had known from birth that 'you can never trust a Campbell'! (I later heard that tragically David Skead had been killed in 1983 in an accident in a small plane on an ornithological assignment. When I met him he had just been promoted into an office job in Pretoria and was fighting having to give up the field work that was his real love. The accident occurred soon after he had managed to wangle himself back into a field job.)

Gaborone was an easy run from Barberspan with the country becoming more arid, and more like Great Sparrow country the further north we went, with a final sharp climb up the escarpment to the frontier. Gaborone is a deceptively small capital as so many of the buildings are hidden by the large trees. We made our way straight to the museum, only to find that Alex Campbell was away in a field trip – his trustworthiness was not going to be put to the test! I asked to see the curator, who turned out to be an immaculately dressed, imposing character behind a large desk. He apologized that he could not help as he did not know anything about birds, nor was there anyone else in the museum in the absence of Alex Campbell who did. On the way out my wife noticed an office with the sign 'Botswana Society' and through the open door an attractive English lady carrying on a vivacious conversation with a visitor. As soon as the visitor had departed we went in and asked for help. In contrast to the curator of the museum, Mrs Cooke was a ball of fire. Yes, there was a Botswana Bird Club, and right away she was on the phone to Mrs Janet Barnes, the chairman. Mrs Barnes had watched a pair of Rufous Sparrows building a nest about four weeks ago and we arranged to see her the following morning when she would take us to the spot. We then turned to Mrs Cooke for some help with accommodation – neither the President nor the Holiday Inn in the capital were our style, but there was a restaurant, the Woodpecker, at the far end of Gaborone Dam about 15 kilometres away where a few rondavels, in the style of the round huts of southern Africa, had recently been put up so that people could stay. This sounded more in our line and we made arrangements through the office in Gaborone. When we got to the Woodpecker there was no sign of life. It was a Monday and all we could find out was that the restaurant was closed on Monday evenings. Mrs Helfer, the proprietrix, is a remarkable lady. A Ghanaian and a practising solicitor, she ran a hairdressing establishment as well as the Woodpecker. We had almost resigned ourselves to spending the

The otters seemed so unconcerned that on occasions we thought they were coming to join us for a drink.

night in the car when at last she turned up, not having heard anything about our arrangement, and all was well. The rondavel was very comfortable and shortly after we had settled in she appeared with an enormous meal. The Woodpecker was an excellent place to stay, alive with birds in the surrounding countryside and with superb food in the restaurant – I am not one of those naturalists who is so obsessed with his prey that he cannot appreciate the good things of life – and a delightful spot beside the dam where one could sit with a drink in the evening watching otters disporting themselves in the water. They seemed so unconcerned about our presence that on occasions we thought they were coming to join us for a drink!

Next morning we called on Janet Barnes to go to look for the Great Sparrows. Before setting off she was most insistent that we should have a cold drink; later we found out that this was at her husband's suggestion so that she could look us over before deciding to disappear with us into the bush. Evidently we passed muster and later when we met her husband I was further able to establish my credentials when we found out that we had both been apprentices at the Metropolitan Vickers Works at Trafford Park in Manchester. Our destination was a small valley in thornbush country some 75 kilometres from Gaborone. We took the tarmac road 50 kilometres west to Molepolole, remarkable in itself as tarmac roads are pretty scarce in Botswana, and we reached the furthest westward extent at Molepolole. We then

Cock Great Sparrow, northern Transvaal.

turned north on a dirt track through increasingly arid country. We stopped at a point that looked no different from anywhere else on the track, left the car, and descended into a hot, but beautiful, valley with scattered acacias and other large trees. After a walk of about thirty minutes Janet located the small thorn tree in which the Great Sparrows had been building their nest. The nest, which had been less than two metres from the ground, had been pulled out, but after a short search I found what was presumably the same pair building a nest in a Karoo acacia, this time over six metres up. The birds were not shy and they were quite unconcerned by my close observations. A search in the valley produced another male Great Sparrow about 700 metres away, but this bird apparently had neither mate nor nest.

The following day it was a long way back again: the tarmac to Molepolole, the hot dusty track to the place we had to dismount and the walk along the Great Sparrow valley to the nesting site. My wife found it a long day under a shady tree doing her tatting while I watched the birds, but having found a nest it was an opportunity not to be missed.

From Botswana we moved on to Naboomspruit in central Transvaal, where we were going to stay on Mr E. A. Galpin's farm, Mosdene. This farm covers about 5400 hectares of acacia thornveld; it was particularly dry and, although there were a number of thunderstorms when we were there, no rain actually fell on the farm. Most of the ponds and waterholes had dried up and the cattle were dependent on water pumped up from deep wells. It seemed to me very marginal cattle country, but looked very suitable for Great Sparrows and that was what interested me. The cottage in which we stayed had no electricity and no fridge, but it did have a paraffin freezer, which was so efficient that it froze the beer in the bottles completely solid; this was most frustrating as when one returned hot and thirsty from sparrow watching there was a long wait before the beer was drinkable. The day after we arrived Warwick Tarboton and Richard Dean, the warden of the nearby Nylsvley Nature Reserve, came to take us round the farm and show us the most likely places to find Great Sparrows. We did in fact find a pair at one of these, but it was in another area close to the cottage where we were staying that I found a pair with a nest. Like the one in Botswana found by Janet Barnes, this was in a small tree only two metres from the ground. Richard Dean was able later to show us another pair with a nest in the Nylsvley Nature Reserve, again in a small thorn only two metres high. He said that there were normally only two pairs in the 6000 hectare reserve, though he considered that only 1000 hectares of this was suitable for Great Sparrows.

At Mosdene I rather disgraced myself by picking up a bug, which resulted in what I called 'lassitude fever'. Luckily, however, this did not prevent me from driving up to the Great Sparrow's nest and watching the birds from the comfort of the car. The birds again appeared quite unconcerned and on one occasion, when we took shelter under a nearby acacia to get out of the heat of the sun, the male actually flew into the tree and worked his way down the branches to look into the car and see what was going on.

These observations indicate that unlike most of the sparrows the Great Sparrow is not a social species. This matched my somewhat limited experience in Kenya where on only one occasion had I seen more than a pair at any one time; this was a flock of about ten birds that were feeding at the Olemelebe Gate to the Masai Mara Reserve in March 1982. Although the birds in southern Africa were well away from human habitations, this is not always the case with Rufous Sparrows. In February 1978 I saw a pair in Lengata, a suburb of Nairobi, and another with a nest in a creeper on an electricity pylon in a street in Nakuru Town; in 1982 I saw a pair in Nyahururu and a single bird in the Masai Mara Reserve came to feed on scraps that we put out at Keekerok Lodge. So perhaps to some extent at least this species does at times show the sparrow characteristic of association with man, though it is unique among the sparrows in its solitary behaviour.

This is a book about birds, but even the most ardent and single-minded sparrow watcher cannot ignore the politics of the countries he visits. News from South Africa is generally bad. One suspects that a lot of this, if not generated by the media, is at least blown up or distorted to increase newspaper sales and television audience ratings. Even in the politically hard-right parts of Transvaal we saw the special 'black' and 'white' entrances to banks and post offices ignored by both colours alike. On many of the farms there seemed to be a good working relationship between the white proprietors and the black farm workers and even one garage was owned by a black proprietor with whom the white staff were much happier than with the previous white owner who had gone bankrupt. It may seem a bit naive, but it appeared to me that the rest of the world, much of which is in no position to throw stones, could do more to help the troubled situation in South Africa by applauding what is good rather than by excessive condemnation of what is bad. I suspect that much of this criticism comes from people who have never been there and have views formed by television and other scarcely objective sources; criticism is more likely to harden attitudes and give comfort to the reactionary elements in the governing minority. All I can say is that I enjoyed my visit, quite apart from my successes with the sparrows, and would return without hesitation. As always, I should never have achieved as much as I did in my sparrow watching without the generous help that I have almost now come to expect from fellow bird enthusiasts.

The early morning trip was slightly marred by the fact that I was bitten by a dog on the way back to the hotel.

10: Sparrow number twenty: The Cape Verdes

On the 16th of January, 1832, we anchored at Porto Praya, in St. Jago, the chief island of the Cape de Verd archipelago.

Charles Darwin, *The Voyage of the Beagle,* 1841

In comparison to the journey out to South Africa to see the Great Sparrow as described in the previous chapter, the return was very peaceful. We had only two hours to wait at Kinshasa and with the confidence inspired by our previous stay did not worry over much about the removal of our tickets and passports, nor the doubts about a connection. I did, however, have one worry on the way home. I was thinking of what I had read about the Cape Verde Rufous Sparrow in a book by David and Mary Bannerman. From their description it appeared to be a social species, occurring in swarms, and, moreover, nesting in inhabited areas as well as away from man out in the countryside. This did not sound like the Rufous Sparrows I had been watching on mainland Africa; by the time we had got home I was already planning our next trip – the Cape Verdes.

The Cape Verde Islands became the Republica de Cabo Verde in 1975, when they achieved independence from Portugal. In order for us to go there it was necessary to obtain visas. This presented the first difficulty. The Cape Verdes has no representation in the UK and the only embassies in Europe are in Lisbon and The Hague. I wrote to Lisbon to get the necesssary application forms and was informed that the granting of visas was a simple process that would only take three weeks once they received the completed application forms. This seemed no problem as it was November and we were not planning to go until the following October. I had not allowed, however, for the reality of the situation and began to get concerned when a week before we were due to depart, despite several follow-up letters, our visas had still not arrived. Fearing that, arriving in a small, newly independent nation by plane from Dakar in Senegal without the required visas, we should immediately be put back on the plane and returned to Senegal, it seemed that desperate measures were called for. I have a friend who has a Brazilian wife and not feeling that I could cope with the situation in Portuguese, even though I had been working away at this for the last year in preparation for the trip, I asked if she would be kind enough to ring up the Foreign Office in Praia, the Cape Verde capital, to try and ensure we should be allowed in. We found the phone number from International Directory Enquiries and then got through to the Foreign Secretary himself at his residence. If somewhat surprised, he was very helpful and transferred the call to one of his staff who assured us that our visas would be ready for us with the immigration officer when we arrived at the airport and told us not to worry. Not being the worrying sort, I didn't, though perhaps in hindsight that was being unduly optimistic. It was a Saturday afternoon when we arrived at Praia in the small plane from Dakar – it had been impossible even to confirm that flight from England, so in getting seats we had crossed another hurdle – and I hung back at the end of the immigration queue so that I could explain the situation in peace without an impatient line of people behind. Of course, there were no visas and the immigration officers had no knowledge of our arrival. They were, however, very understanding (of the situation, not my Portuguese!) and let us in provided we agreed to attend at the immigration office on the Monday morning. As surety our passports were confiscated.

We had arrived! And having settled into our hotel immediately went out to look for the local Rufous Sparrows, just in case we should be deported after a couple of days. It is a desperately arid place and a short walk revealed very few birds, the only sparrows we saw being Willow, not the one I was looking for, the Iago Sparrow: the Rufous Sparrow in the Cape Verdes is called the Iago Sparrow after São Tiago, the main island in the group where Darwin had collected it in 1832. The following morning I was out at dawn and after passing the local joggers – how had this bit of recent Western culture got to the isolated Cape Verdes? – soon found a hen Iago Sparrow feeding young in a nest in a hole in a lava cliff. We found others during the Sunday so, even if our visit was going to be cut short the following day, at least I had seen the bird.

Before going into Praia on the Monday, hopefully to obtain our visas, I went out again at dawn to have another look at the nest I had discovered the day before. This early morning trip was slightly marred by the fact that I was bitten by a dog on the way back to the hotel. Outside the UK I am slightly nervous about dog bites, but Gloria, the American owner – she was the wife of one of the US embassy staff – assured me that the dog had had its anti-rabies injections and that I was not at risk. As further reassurance she told me that the dog had already bitten the ambassador's son and that he had not succumbed. What the ambassador had said to Gloria's husband was not recounted!

This episode actually turned out to our benefit. The Visa office was a window in a busy street in Praia. When we arrived at 9 o'clock it already resembled a rugby scrum with a pack of Caboverdeans trying to get exit visas to go to America where a large proportion of the population emigrate in order to find work. Not knowing the ropes and knowing little of the language, we had little hope. It took some time to find out that first of all we had to get some fiscal stamps. That seemed easy, but the fiscal stamp office was closed because it was the end of the month and they were balancing the books. We failed to find the agency where we were told the stamps were being sold. I went to the US embassy where I knew I could be understood to see if I could get some information – there is no British embassy in the Cape Verdes, the UK being represented from the Ivory Coast or some other equally inaccessible spot on the west coast of Africa. Fortunately, as the recipient of a bite from an American dog, I needed no introduction and was passed on to a Caboverdean Mr Fixit. He took us back to the fiscal stamp office at the other end of the town – fortunately Praia is a small capital – naturally it was still closed, but he had not believed me when I had told him this; however, he knew the location of the window where we could get the stamps. We had passed it before and I should have identified it quite easily from the long queue outside. We were prepared to join the end of the queue, but he was not. A little money and a chat at the head of the queue and we quickly had our stamps. Naturally by this time the visa office was closed for lunch and the afternoon siesta. When it finally opened we were told our passports and visas would be ready at 5 p.m. It had been a traumatic day – what it would have been like without our guide and queue jumping I dread to think – but we were legally present and ready to get down to planning our stay over some much needed drinks.

There are ten inhabited islands in the archipelago, mostly accessible by plane from São Tiago, and we decided to visit four: São Vicente and Santo Antão lying at the west of the Windward Group, and Fogo in addition to São

Tiago in the Leeward Group. São Vicente was chosen because it was the only island on which the House Sparrow is present, Fogo because it was the one island on which the Iago Sparrow was reputed to be absent.

We fixed a flight to São Vicente on the Wednesday and proceeded to do some preliminary exploration of São Tiago. Unfortunately, my wife took ill on the Tuesday and, at the sight of a plate of chips in the dining room that evening, passed out. Although she felt better the following morning, she thought it wise to miss breakfast and by midday she felt fit enough to take the flight to São Vicente. It was over two weeks before we returned to the hotel where the incident had occurred. When she walked into the dining room that evening, there were evident signs of relief on the faces of the long-term residents. They had assumed that she had contracted some fatal Caboverdean bug and by this time was safely interred.

The Cape Verdes present a pretty harsh environment and support only a very limited avifauna. Excluding the seabirds, there are less than thirty species, with four or five of these probably recent introductions, but among the thirteen songbirds (probably now only twelve as the introduced Goldfinch appeared to have died out) no less than three sparrows. What a place for the sparrow watcher! I have already discussed the Willow and House Sparrows in Chapter 5; here I want to concentrate on the Iago Sparrow that was the real reason for the visit.

We found the Iago Sparrow to be widespread on São Tiago, São Vicente and Santo Antão, breeding in holes in buildings in inhabited places as well as in lava cliffs, both on the sea coast and inland, and under rocks in the barest flat ground. Not only were they breeding in close association with man, unlike the Rufous Sparrows on mainland Africa, but they were quite social – no less than eight pairs were nesting under tiles on the roof of the small Baptista Sousa Hospital in the central square of Mindelo, the one town on São Vicente, where I could comfortably watch them through a telescope from the balcony of our hotel bedroom. Most of the nests we found were in holes in lava cliffs. Going in and out of the holes in the rough rock is pretty hard on the birds and the ones that were breeding could be readily distinguished by the tatty state of their feathers. A complete moult and a new set of feathers after the breeding season is obviously of paramount importance if the birds are to retain their power of flight. Maintaining the feathers in perfect condition is essential to the survival of a bird; the birds nesting in the lava cliffs could well have been restricted to one brood per year for this reason. Most of the other species of sparrows are multi-brooded.

Rainfall is pretty scarce and irregular in these islands – we were told that it was over seventeen years since it had been possible to raise a crop of maize and that all grain had to be imported. In 1983, 80 mm of rain had fallen on São Tiago on 25–26 August and a further 100–300 mm in different parts of the

Cock Iago Sparrow: sparrow number twenty.

island on 13 September, when it had rained all day. On our arrival at the beginning of October there was some emergent ground vegetation and numerous brightly coloured caterpillars and grasshoppers, sufficient to have triggered off the breeding of the Iago Sparrows that depend on animal food for rearing their young. When we returned to São Tiago on 17 October, the vegetation had already died, the maize was in ear, but brown and not likely to survive without further rain, the caterpillars had disappeared, though grasshoppers were still there in abundance, enough to satisfy the Iago Sparrows. On São Vicente, which is even drier than São Tiago – not only had there been no rain this year, but according to a local resident there had been no significant rain for the past sixteen years – we found the Iago Sparrows breeding only in Mindelo, where at least there was some green vegetation in the watered gardens, the rest of the island including the golf course being almost entirely bare sand with a few rather sorry looking acacias. The House Sparrows that were also present in Mindelo were not breeding and the Willow Sparrow seemed to be on the way out as we only saw hybrids with House Sparrows. Willow Sparrows had already died out on Brava, one of the other islands, in the prolonged droughts of the 1940s, though they have recolonized since our visit following rather moister conditions. House

Sparrows were not numerous, but they had survived there since 1924; presumably they must breed at some other time of the year.

An unexplained puzzle is that the Iago Sparrow has never been recorded from Fogo and there the Willow Sparrow reigns supreme. We flew direct from São Vicente to Fogo and when we arrived a German lady kindly offered to give us a lift to our hotel – there was no taxi. She was the wife of Peter Sauer who was studying the hydrography of the island under a German-financed agricultural project. There are several springs on the island, all at sea level. Peter Sauer told us that when rain fell in the central mountains it took seven years to percolate the porous volcanic rock before it emerged from the springs, from where it had then to be pumped back up to irrigate the fields. The hotel in which we stayed in São Filipe had been built about seven years before. It had a large swimming pool advertised on the brochure, but this had never been filled as there was not enough fresh water and it was too high above the sea for salt water to be pumped up. The frogs were enjoying an inch or so of water at one end, but were very vulnerable there to the attractive Grey-headed Kingfishers; it may seem odd that there is a kingfisher in such a dry place with no standing water, but this is largely a dry-land kingfisher and feeds mainly on the numerous small lizards. The first room we were offered overlooked the courtyard and had no balcony. We moved to another with a balcony overlooking the sea, but no wardrobe. We gave up. However, the

The frogs were very vulnerable to the attractive Grey-Headed Kingfishers.

following day we were given another, complete this time with wardrobe and a private balcony that had a good view of several Willow Sparrow nests in some beautiful flowering mimosas. All parties were now satisfied.

There was a limited amount for Astrid Sauer to do, apart from reading, and I suspect that time hung rather heavily on her hands. She very kindly not only picked us up at the airport when we arrived and took us back on our departure, but also ran us about in between and gave us a lot of interesting information about the place. Fogo is unbelievably dry for an inhabited island; when we were there no rain had fallen for the past seventeen years. It is a volcanic island that erupted as recently as 1951 and a wisp of smoke was still coming out of the top. The cone rises to 2830 metres and is high enough to attract clouds and, even if rain rarely falls, there is sufficient condensation on the walls of the caldera to provide enough water in the crater for some cultivation. We were surprised to find familiar European vegetables such as lettuce, cabbage and peas. Not even these can be grown on São Vicente and the vegetables have to be brought there daily by ferry from Santo Antão, which is slightly less arid. There were also apples, and to our even greater surprise some vines and a small wine-making industry, enough to make about 50,000 litres in a good year. This is a full-bodied red wine and we were able to buy a bottle for 75 CV Esc. (about 70p). There is every reaon to believe that the volcano is still active and the caldera seems a risky place, but the fact that it provided a living induced quite a significant community to live there.

Willow Sparrows were all over Fogo, breeding in trees round the hotel, in a small clinic in São Filipe where the patients could not understand what I was doing skulking in the shade and watching the roof through binoculars, and right up to the walls of the caldera at 1650 metres. We saw nothing of Iago Sparrows and, despite odd reports, I do not think they have ever occurred there since the first ornithological records that date back to 1898; none of the local residents I spoke to could ever recall having seen one, though they knew the bird from the other islands.

Leaving Fogo, where from the air the lava-flows from the eruption could be seen vividly running down the side of the mountain, we returned to São Tiago. This time, after a certain amount of difficulty, we were able to rent a car – a deposit of £50 and no questions, but an instruction that if we were stopped by the police we were to say that it had been lent to us by Toni (surname unspecified) whom we had met in a bar. The problem at the car rental garage was that all the cars were standing idle with no engines – spares are difficult in the Cape Verdes. Our first job was to reconfirm our flight back home, which by international air regulations must be done 72 hours prior to departure to avoid cancellation. At the TACV (Transportes Aéreos de Cabo Verde) office we were told that our flight from Dakar to Paris and on to

London was OK, but there were no seats available to Dakar and no way we could get there in time. After a heated conversation wih the girl at the desk in my limited Portuguese and a long telephone conversation on her part, two seats became available for the flight I had originally booked.

Having established our escape route home, we started to explore the island. Tarrafal at the north tip had been a penal colony in the days of the Portuguese occupation. Heading north we stopped for a beer in Santiago and confirmed that there was petrol to be had at Tarrafal. Tarrafal seemed to have changed little since the days of the penal colony. It appeared completely deserted and the petrol station had clearly not dispensed anything for many years. It took some time to find an inhabitant; we asked about petrol and he pointed to a small village about three kilometres along the coast. Driving there we found little apart from a store and no sign of petrol pumps. Behind the store, however, there were some 200-litre fuel barrels on stillages. The next problem was that none of the keys on the key ring would unlock the petrol cap, nor would even one from an enormous bunch of keys produced at the store. Eventually we found that the lock on the cap was completely rusted up and no longer functioned; with a little effort we were able to unscrew the cap. By the time we had topped up laboriously with a jug, the store had

Petrol station, Tarrafal, São Tiago, Cape Verdes.

closed and I had to go round the back to pay. The big family was sitting in the open round a large table with an enormous fish and bowls of rice. We were invited to join them. Such were the local people. Not like the bureaucrats that took a day to sort out our visa problem. It did not do to have car trouble on the São Tiago roads. On the way back to Santiago, something over 50 kilometres, we passed exactly three vehicles, two cars and one lorry.

This had been an exciting trip for me. I had found the Iago Sparrow and found it to be very unlike the Rufous Sparrows I had watched on the African mainland. The Cape Verde birds were not only social, nesting together in colonies, but were also living in close association with man. These and other features led me to the conclusion that here was a different species and I had increased my searches from nineteen to twenty. Like the Great Sparrows in South Africa they were very tame and allowed me to approach as close as 4 metres to take photographs, as they scuttled about on the ground like small mice, but apart from their similarity in plumage in no other way did they look to me like Rufous Sparrows.

Our improbable story of being in pursuit of a sparrow seemed to satisfy the sentries of our good intentions.

11: A success story: Israel and Cyprus

This bird is the most limited in its range and the scarcity in numbers of individuals.
Canon H. B. Tristram, *The Survey of West Palestine,* 1884

Canon Henry Baker Tristram spent much time in the Middle East and North Africa for the good of his health. While the climate may have suited him, his presence there was not good for the health of the birds – Tristram was an avid collector. He discovered a number of new species, including a grackle and a warbler that bear his name. The bird that is associated with him that interests me is a sparrow that was collected by R. M. Upcher in 1864 from the southern and southwestern end of the Dead Sea and named by Tristram as the Dead Sea Sparrow. Tristram considered this to be one of the rarest and most restricted species in the world and even fifty years later this was still the situation in Palestine, as confirmed by Carruthers, the same indefatigable traveller we first met in Soviet Central Asia (see Chapter 2). However, in 1888 a population of this bird had been discovered in Sistan, the delta of the Helmand River in the desert region lying across the frontier between Iran and Afghanistan, and yet another in 1904 on the lower reaches of the Karun River in southern Iran.

By 1919 it had spread to the Jordan Valley at the north end of the Dead Sea and in the same year it was found near Baghdad on the River Tigris, though at this date still considered a rare bird. Since 1950, however, there has been a dramatic spread over much of Israel, new colonies have been reported from Turkey, and in 1976 it arrived in Cyprus.

This was for me a fascinating situation and obviously one that had to be investigated on the ground. Israel seemed to be the obvious place to go after it; not only was it discovered there and takes its name from the Dead Sea, but it is the country where it has been most studied and a place where there are few travel restrictions. I had previously been in touch with Professor Heinrich Mendelssohn, Emeritus Professor of Zoology at Tel Aviv University, and Professor Yoram Yom-Tov, the current incumbent of the chair, regarding their studies on the bird, and on learning that we were coming to look for it Yoram Yom-Tov very kindly suggested that we should spend the first night with him so that he could brief me on the best places to go.

We were most hospitably met when we arrived at 2 a.m. – the aeroplane is quite the most socially unacceptable means of travel. Suitably rested, we got down to details of planning our itinerary. This was October and thus outside the breeding season and, as Mendelssohn's studies had largely centred on breeding distribution and Yom-Tov's on breeding biology, it seemed wise to cover a number of different places in case the birds had dispersed from any of the breeding localities. The plan was to stay for a few days at the field study centre at Hazeva, where we could visit the area at the southwest end of the Dead Sea, where it was first reported, and the kibbutz at Ein Yahov, one of the sites where Yoram had carried out his breeding biology studies. From Hazeva we were to go to Elat, where we were most kindly offered the use of a university flat, and although this is an area where the bird had only recently begun to breed it was important nevertheless as a place where it was regularly seen in the winter months. We could then go north to the Beit Shean Valley, south of the Sea of Galilee, the second of Yoram's study areas and also where we could visit Shemuel Lulav, the curator of the natural history musuem at Beth Gordon and another man who had been working on Dead Sea Sparrows. Yoram had nominated suitable contacts at all these places and with all this expert help we expected no problems.

The first minor difficulty arose the next day when we were picking up the hire car that I had ordered in Tel Aviv. While I was signing the papers for the car, I dispatched my wife to the nearest bank to get some *shekalim*. She came back empty handed: the French franc had just been devalued and dealing in foreign currency had been suspended. We had brought enough Israeli money from home to cover our immediate needs of food and petrol so we just pressed on. Money, in fact, became a major problem: there was nowhere near Hazeva that we could change traveller's cheques and moreover by then it was

Yom Kippur so that there was no point in going to Elat, the nearest place we could be certain of finding a bank, as all the banks would have been closed. The first real monetary problem came when we had to settle our bill at Hazeva. This was on our return route from Elat on our way north and my suggestion that we should defer payment until we passed through again in a week's time was accepted without demur, even our request that we wanted to borrow our bed linen, as there was none at the flat in Elat, was acceded to with a smile. Birdwatchers must enjoy an enviable reputation for honesty and trustworthiness!

In fact when we finally reached Elat it turned out to be a Friday and the banks were again closed as they do not open on the eve of the Sabbath. Things were not much better the following week as this was Succot, during which there were further religious holidays when the banks were again closed for business, not only on the holiday itself, but also on the previous day as well. By keeping our eyes open, however, we did manage to find one of the short periods when the banks were open and replenish our diminished stock of *shekalim*.

Back to the sparrows. Our first sortie was to Ein Yahov where we quickly located our contact, Gideon Kenaani. He told us that one of Yoram Yom-Tov's colonies no longer existed. This had been in a peach orchard, but the trees had been felled as the fruit did not ripen sufficiently early before the Italian ones for them to be competitive in the European market so that they had had to be sacrificed for something more profitable. He did, however, show us the site of another colony in a group of tamarisks, but remarked that it would be unlikely that we should see any birds as they disappeared after the breeding season. He was right! We found forty nests, but no sign of any birds.

We then visited the sites at the southern end of the Dead Sea, where the bird had first been discovered, but again with the same result. This was the day on which President Sadat of Egypt had been assassinated and the numerous military checkpoints on the road were stopping every vehicle. We had left our passports at Hazeva and once again were in trouble. But not for long, as our improbable story of being in pursuit of a sparrow seemed to satisfy the sentries of our good intentions. This obviously made us well known at all the checkpoints and on subsequent journeys the eccentric birdwatchers were always saluted and waved through without further ceremony.

Having failed to locate Dead Sea Sparrows at either of these locations, it seemed likely that they must have moved south to Elat, where we knew that migrants had been trapped at the bird-ringing station located on a kibbutz just to the north of the town and where the bird was described in the reports of the ringing station as a common winter visitor and rather numerous passage migrant. That assumption was not justified. There were numerous

Great Grey Shrike: the only occupant of a deserted Dead Sea Sparrow breeding colony.

House and Willow Sparrows in the ringing area, but nothing could be found of the elusive Dead Sea Sparrow. We found that the ringing station was only manned in the spring and discussions with our local contact, thoughtfully arranged by Yoram, revealed that the Dead Sea Sparrow did not normally turn up in Elat until December.

Elat, however, did provide a number of interesting compensations. One of these was the underwater observatory on the coral reef just off the coast. This is a cylindrical chamber, with glass windows round the complete circumference, into which one descends down an easy staircase to a depth 4.5 metres below sea level. Here one can watch with great enjoyment a panorama of the colourful denizens of a coral reef from a comfortable room without the anxiety of immediate death by drowning that always accompanies my snorkelling activities. We were further privileged by having as an expert guide Mrs Popper from the marine laboratory that the University of Tel Aviv operates at Elat. Even better, accompanied by Mrs Popper we were allowed in before the official opening time and had the observation chamber to ourselves for an hour. Understandably it is a popular place to visit and to get there before the crowds turn up is an enviable experience. Also, this was the time just before Israel handed back the Sinai to Egypt so that we were able to make a short trip into the interesting desert region there; I never pass the opportunity of visiting a desert region.

About eight kilometres north of Elat lies the Hai-Bar Arawa Biblical Reserve. This was founded in 1968 as part of a chain of reserves set up to re-establish animals that were indigenous in that part of the world in biblical times but have since disappeared. The Arawa reserve consists of over 3000 hectares of semi-arid country enclosed by a strong fence in which there are free-ranging breeding herds of some six species of ungulates – including the wild ass, the white oryx and the addax which are endangered species worldwide – plus four naturally occurring predators. With luck these can be seen in their natural habitat as one drives leisurely round in one's car. The chief curator at the time of our visit was a remarkable American, Bill Clark, whose dedication to his animals matches in well with this establishment that is entirely maintained by voluntary contributions. Bill, of mixed American Indian and Irish parentage, is a delightful companion and a man of many parts as we found when we visited him that evening in his home in Elat surrounded by an extremely catholic library and listening to a Mozart concerto. Israel is a land full of enthusiasts – most of them enthusiasts for their country – but Bill Clark, an American, is enthusiastic for his animals and the breeding project. Arawa did provide sparrows, but again only House and Willow.

It was time to move north again, back through Hazeva where we were able to settle our debt and return the bed linen. On the way I stopped to show my wife some Chukar, the local partridge that she had not seen up to then, but instead she drew my attention to a much larger bird a little way off the road. If I had not stopped for the Chukar we should have missed the Houbara Bustard, the only one we saw on the trip. After Hazeva, where the warden apparently had had no doubt that she would see us again to repay our debt and return the linen, it was back again along the west coast of the Dead Sea, with an eye open for sparrows, to Jerusalem for the mandatory bit of sightseeing. On the way we made a detour to Yitav, north of the ancient city of Jericho, another reported site for the Dead Sea Sparrow, but once more without success.

From Jerusalem, the next objective was the kibbutz of Dede Eliyyahu in the Beit Shean Valley, Yoram Yom-Tov's other study area. We failed to find our contact there, but he soon located us searching the agricultural land and was in no doubt that we were the sparrow watchers about whom he had been forewarned. Again it was the same story; the Dead Sea Sparrows had not been seen there since the breeding season. We were kindly invited by our guide to his house to meet his wife and have some tea, though it was made

I stopped to show my wife some Chukar, but instead she drew my attention to a much larger bird.

abundantly clear that, being a Friday, we should have to leave promptly at 4.30 in order to allow the family time to carry out the prescribed ritual preparations in time for the Sabbath. Our next port of call was the A. D. Gordon Agricultural and Nature Study Institute at the Deganya 'A' Kibbutz at the southern end of the Sea of Galilee, where we wanted to see Shemuel Lulav, the curator of the museum and another of the Israeli Dead Sea Sparrow enthusiasts. Lulav at last was able to show us some Dead Sea Sparrows, even if they were only stuffed ones on display in the museum, and also to suggest a number of places we might find them in the field. We took as our base another field centre, at Qazrin in the Golan Heights. This was in one of the new Israeli towns built in this recently occupied area. Even if a number of the houses had not yet been occupied by human beings, plenty of House Sparrows had moved in.

We spent the next few days searching the various localities near the Sea of Galilee suggested by Shemuel Lulav, but with the same old story and increasing doubts about the existence of the bird at all, until it was our time to return to Tel Aviv on our way home. There was one last place to visit on the way back – the nature reserve at Huleh, the first reserve of many to be established in this small country. We could not have chosen a worse day. It was a Saturday – a holiday for many in this country that contains a number of non-practising Jews. To make matters worse, this was the first day that a new walkway over the marsh had been opened and all the world was there with his wife and children. The place resembled a fairground more than a nature reserve and I had finally accepted that a return visit to Israel in the breeding season was going to be necessary. We were sitting in a large observatory tower among crowds of people, when a young warden came and sat down beside us. With binoculars, telescope and camera with telephoto lens he must have realized that we were more seriously interested in the birds than most of the other visitors. He asked if there was any bird we particularly wanted to see. I had given up, but my wife is not so easily put off, and she told him that we were looking for Dead Sea Sparrows. He had never heard of the species – to be fair he was only there on temporary duty during a leave from army service to help out with the expected crush that day – but he said that if we would wait he would get in touch with the other wardens who were in radio contact. In twenty minutes he was back with the news that there was a flock of Dead Sea Sparrows on a part of the reserve not normally open to the public. He would take us there, but could not leave his wardening duties until four o'clock when the reserve closed and there would be some peace. We were due back at Tel Aviv that evening to stay with the Yom-Tovs. After a telephone call to Yoram he fully appreciated the situation and all we had to do was to find a bed for the night before returning to the reserve promptly at four o'clock. Yuval Erhlich was waiting for us. He unlocked one of the gates and we drove into the reserve along a grassy track above the marsh; we stopped at another gate where we dismounted and I was instructed to go ahead cautiously. I had only gone about 100 metres when I came across the birds; there was a flock of between 60 and 70 of them feeding on the ground on the seeds of small plants. I was able to watch them until dusk and then made arrangements to return at dawn the following morning. Again they were in the same place, feeding on the path as before and this time also on the seedheads of papyrus and tamarisk. I spent another two happy hours with them before having to pack up and return to Tel Aviv in time for lunch with the two professors at the University. I was fascinated to find out during conversation that Professor Mendelssohn was a direct descendant of the composer. Being a Jew, Jacob Ludwig Felix Mendelssohn-Bartholdy, to give the composer his proper name, could not be recognized in Nazi Germany; his popular music, however, continued to be played there during the late 1930s, but was attributed to an unknown composer, it being impossible to ascribe such music to a Jew!

The male Dead Sea Sparrow is a dapper little bird with a small black bib, two yellow spots on the sides of the throat, and a prominent chestnut scapular patch on the wings. It is significantly smaller than the House Sparrow and builds a surprisingly large nest for a bird weighing only 15 grams. The nests seen at Ein Yahov were in tamarisks, built of dead tamarisk twigs, 300–450 mm deep and 300 mm in diameter with an entrance spiralling down from

Professors Yom-Tov and Mendelssohn: my Dead Sea Sparrow mentors.

the top. This gives an open structure that allows air to circulate through and prevents excessive heating of the eggs by insolation.

In Israel the Dead Sea Sparrow now breeds from the Arawa Valley south of the Dead Sea to north of the Sea of Galilee and west along the valleys of Esdralon and Zevoulun almost up to the Mediterranean coast. In its original habitat, in some of the hottest parts of the world, the hen did not incubate the eggs during the day and depended on the presence of water to keep herself cool and prevent overheating of the eggs. Mendelssohn and Yom-Tov attribute the extension of range in Israel to increased availability of food in the form of weed seeds in the cultivated kibbutzim together with the creation of fishponds that provide open water, plus a change in breeding behaviour whereby the eggs are incubated in the normal way in the cooler areas that it now inhabits. This is a fascinating turnaround for a species that from its disjunct distribution was looking like a relict that was on its way to extinction.

I was left with the question of what happens to the birds after the breeding season. Considering the large numbers that now breed in Israel it is surprising that we did not come across some when searching the country. My first thought was that a large proportion might emigrate and that the birds that occurred in Elat in the early spring were on their way back from winter quarters further to the south. A possible place would be the Arabian peninsula, but M. C. Jennings, who has recently published a check-list of Saudi Arabian birds, informed me that he had no evidence of any occurrence of Dead Sea Sparrows in that country, though the west was admittedly infrequently visited by birdwatchers. The same applies to the countries on the west of the Red Sea – lower Egypt, Sudan and Ethiopia – though John Ash who made an extensive study of the birds of Ethiopia in the early 1970s did not see any there. Another possibility is that the birds disperse into a different habitat, dispensing with ready access to water that they require for breeding, and perhaps I was looking for them in the wrong places. This would be an interesting problem for someone to solve.

Whatever one feels about the politics it is impossible not to be impressed about what has happened in Israel. There is a real excitement about belonging to the country that I have not experienced elsewhere. And their achievements are many. Nowhere is this more impressive than the view from the side of the Rift Valley where the border between Jordan and Israel is drawn sharp and clear: brown in Jordan, green in Israel, despite no difference in geology. The secret is that the Israelis separated the sweet and saline water from the wells and springs, using the former for irrigation and the latter for fishponds that must give a significant protein yield. It is a small country as one is constantly reminded by coming up against frontiers – bad news for birdwatchers and photographers, who are normally looked at askance in military zones. Not so, however, in Israel. I found the soldiers most friendly and courteous, provided one took the elementary precaution of not actually photographing military installations.

The Dead Sea Sparrow was first proved to breed in Cyprus at the Akrotiri Salt Lake in 1980, though it is pretty certain from information that I was able to obtain from Peter Stewart, one of the authors of a recent list of Cyprus birds, that they first arrived in 1976. As I had been unable to make proper observations on them in Israel it seemed a good idea to go there, not only to see them at a breeding colony, but also at their first breeding site in Europe. This we did in 1985 and found 35 nests in tamarisks growing in the edge of the lake. I was there in April, just before the breeding season was fully under way. The nests are very soundly made and stay in position for a number of years, so the number of nests does not necessarily represent the exact size of the breeding colony; neverthless there was a good number of birds around defending the nests and adding new material. I spent several happy hours standing in the soft mud of the lake with water up to my knees, not as salty as the Dead Sea, but with that oily feel and very warm so that it was not unpleasant. I suspect that birds spread to Cyprus from Turkey, where they were first recorded in 1964 and have since spread to the west in a similar way to that in Israel. One mystery, however, remains; like the birds in Israel both the Turkish birds and the Cyprus ones disappear after the breeding season to as-yet-unknown winter quarters.

We had one rather odd experience in Cyprus. On a day off from the sparrows, we went up to the Troödos Mountains. Driving back from the radar station I saw what looked like a wallet lying in the middle of the road. A wallet it was, and it contained an American passport, a Saudi Arabia work permit, American Express traveller's cheques to a value of about $3000, and a business card from a car-hire firm in Limassol. On returning to Limassol I rang up the American Express Office to cancel the cheques, but it was a quarter to five on a Friday evening, and they showed little interest, suggesting I call back on the Monday morning. At the car-hire office I was able to confirm that the owner of the wallet had in fact hired a car and was due to return on the Monday. I left my address and a message suggesting that he got in contact. There was also a hotel receipt in the wallet and I left a message there. That seemed as much as we could do, so we returned to our apartment, changed and went out for dinner. Later that evening, after my wife had gone to bed and I was enjoying a last brandy on the balcony, there was a knock at the door. It was the owner of the wallet. He came in and when I found out he was an engineer we had a lot in common. We had a long chat and when my bottle of brandy was finished he departed. Not even a word of thanks!

With the Cinnamon Sparrows rather upmarket at the larger houses with gardens and open spaces . . .

12: Moving east: The Himalayas

Ship me somewhere east of Suez.
Rudyard Kipling, '*Mandalay*', 1892

The next species on my list was the Cinnamon Sparrow. This is a high country species and my first thought was Nepal in view of its accessibility to Western tourists. According to the *Birds of Nepal* by Fleming, Fleming and Bangdel, it is common in the northern Dolpo. Failing to find this region in any of my extensive collection of atlases, I wrote to the Flemings to enquire if the Dolpo was sufficiently accessible to Kathmandu for it to be possible to make useful observations on a three-week visit. The reply was that not only was the Dolpo a three-week trek from Kathmandu, but as it was a sensitive border area it was doubtful whether a foreigner would be given permission to visit there. Instead, however, Fleming suggested Dhanaulti in the Garhwal Himalayas close to the hill station of Mussoorie where he had found Cinnamon Sparrows reasonably common a few years ago. Attention was thus turned to India for a visit in 1981, but it proved difficult from this end to make the necessary detailed travel arrangements and I contacted the Delhi office of the company I worked for to see if they could help. This resulted in a delightful friendship with Sugato Chaudhuri, who not only made the arrangements that I asked for, but met us at Delhi Airport at 4.00 a.m., two hours later than our expected arrival, took us to his home for some rest, and then provided a network of contacts wherever we went in India to ensure that if we should encounter any difficulty there would be someone on the spot we could turn to for help. This was another of those happy spin-offs from sparrow watching. Since then I have spent many happy hours with Sugato and his delightful family; they could not have done more to make my visits to India both rewarding and enjoyable. I am pleased in turn to have been able to welcome several of them to my home in England.

Everyone we met in India on this visit asked without fail why we had come in May, the hottest time of the year. It was of course because I wished to see the birds during the breeding season, but the enquiry was understandable. When we were travelling we usually set off about 5.30 a.m. in an effort to avoid the worst of the heat, but even by 10.30 the air coming into the car was becoming uncomfortably warm as well as very dusty. The move was then to shut the windows until the temperature inside became unbearable, when the windows came down again. From then on the journey lapsed into an alternate opening and shutting of the windows as we debated which was the lesser of the two evils. We usually carried an icebox with cold drinks in the boot. It is a feature of India that wherever you stop, even in the remotest parts with no sign of life, within minutes an audience crowds round. The teeming population remains my greatest impression of the subcontinent.

Mussoorie is the closest of the hill stations to Delhi. We went by car to Dehra Dun, 250 kilometres north of Delhi, and then to Mussoorie, only 35 kilometres away by a dramatic climbing road, but at over 2000 metres nearly 1400 metres higher and of course much cooler. On arrival at our hotel, Hakman's Grand Palace, I went out on to the rickety veranda outside our room and the first bird I saw was a Cinnamon Sparrow. This was, however, the only attraction of the hotel, which was very run-down and dirty; whoever Hakman was I do not think he would be very flattered to have his name still associated with the hotel, which was neither grand nor palatial. I must hasten to add that this had been booked at my request by Sugato Chaudhuri, following a recommendation from the Indian Tourist Office in London and no blame can be attached to him; all the arrangements he made were flawless. In fact, he recovered the situation by proxy: we got in touch with his local back-up contact – a school friend he had not seen since his schooldays – and quickly transferred to the small Roselyn Estate Hotel most efficiently run by Prem Kishinchand Thadhani and his wife, though it did lack Cinnamon Sparrows in the grounds.

I found a total of 50 pairs of Cinnamon Sparrows in Mussoorie and the adjoining town of Landour, where Sir George Everest, the first Surveyor General of India and after whom the mountain was named, lived from 1833

A handsome Cock Cinnamon Sparrow, Mussoorie, Garhwal Himalayas.

to 1843 with his Indian mistress safely tucked away in a house at the bottom of the garden. House Sparrows were also present in Mussoorie, occupying the closely built-up area round the library and the crowded Kulri and Landour Bazaars, with the Cinnamon Sparrows rather up market at the larger houses with gardens and open spaces round them. The Cinnamon Sparrows' nests were almost equally divided between holes in trees and typical House Sparrow sites under house eaves; one nest was even in an electricity junction box on a lamp standard, a site I have seen used by House Sparrows in many parts of the world.

Mussoorie is a popular holiday centre with the visitors spending much of their time walking up and down the Mall with the hillside falling steeply to the Doon Valley and a magnificent view over the plains of northern India. I found a Cinnamon Sparrow's nest in a hole in a chestnut tree at the side of the Mall and was able to advance my camera on a tripod gradually to within three metres without disturbing the parent birds from continuing to feed their young. This was a constant source of amazement to the continuous stream of passers-by who could not understand why I was photographing the view of the valley below with a tree obscuring most of the scene nor could they resist pointing out a better place for me to take my photographs.

Passers-by could not understand why I was photographing the valley below with a tree obscuring most of the view.

From Mussoorie we went along the range of hills past steep terrace cultivation and flowering rhododendron trees to the small, stone-built village of Dhanaulti, situated among enormous pines and deodars 24 kilometres away. It had not been necessary to follow Robert Fleming's advice to go there to find my Cinnamon Sparrows, but it was a good opportunity to have a further look at the area. Cinnamon Sparrows were regular in the cultivated land along the way, both in the terrace cultivation, where they were nesting in holes in rhododendron trees, and at the little wayside hamlets, where they were nesting in holes in the walls of the stone-built houses; one such nest was in a hole only one metre above ground level. There were six pairs in Dhanaulti, with its 20–30 stone houses. Dhanaulti is about 250 metres higher than Mussoorie and by the middle of May there was no sign of breeding having started, whereas in Mussoorie most pairs were already feeding young in their nests. One pair was seen disputing a hole with a pair of Black-crested Tits which were feeding young inside. The much smaller tits evidently came out on top, for the next day they were feeding their young in undisputed possession of the hole.

I was fortunate to have a business trip the following December that took me to the Indian Institute of Petroleum just outside Dehra Dun. I managed on the Saturday afternoon to revisit Mussoorie. It was a very different place than in May with a cold wind and overcast skies. The House Sparrows were still present in the same places where I had seen them before, but there was no trace of the Cinnamon Sparrows. One of the few birds I did see was a brilliant scarlet and black Wallcreeper that was working its way along one of the stone ditches at the side of the street, a bird that I had previously looked for unsuccessfully in many remote and difficult mountainous regions. The following morning, however, I found a flock of about seventy Cinnamon Sparrows beside a small river 10 kilometres south of Dehra Dun, probably only 30 kilometres from Mussoorie as the sparrow flies, but 1400 metres lower in altitude and 20°C warmer at that time of year. There are reports from other parts of its range that the Cinnamon Sparrow makes short vertical migrations in the winter. So, although a montane breeder, it does not seem to be as tolerant of the cold as the versatile House and Tree Sparrows that remain in some places at high altitudes throughout the year, possibly helped by their closer association with man in being able to find food.

This visit to Dehra Dun was made luxuriously in the comfort of a diplomatic car. The effect of a CD plate is quite remarkable: police held up the traffic to let us through and checkpoints were passed with a wave of the hand. This is a tremendous boon on the extraordinarily crowded Indian roads. I was only disappointed that this privilege was granted me as a visiting industrial lecturer, not for my reputation as an international sparrow-watcher.

The hill station at Mussoorie was only established in 1827 so it would appear that the House Sparrow, dependent as it is on human habitation, can only have been a comparatively recent arrival, no doubt preceded by the Cinnamon Sparrow that presumably arrived with the advent of terrace cultivation to that area. The dating of this has not been precisely established, though Dr Ashok K. Ghosh of the Department of Anthropology in the University of Calcutta has suggested to me that it probably began between 4000 and 3000 years ago. This would make even the Cinnamon Sparrow a comparative newcomer; no doubt it was the first sparrow to become established in the Garhwal, though clearly this could not have been the region in which the species first evolved.

House Sparrows were not present in Dhanaulti, though they were common at Chamba, the first town we passed through on descending from the hills. This is an old-established town, no doubt occupied by House Sparrows for a long time. House Sparrows were also present at Thatur, a small village about 15 kilometres north of Mussoorie, but at only 1200 metres presumably too low for Cinnamon Sparrows.

Lest it not be thought that I am completely indifferent to culture we could not come home without visiting the Taj Mahal.

13: Two birds with one stone: Land of the Five Rivers

In regard to this supposed species I am fast verging on Betsy Trigg's conviction in regard to Mrs Harris, and if such a bird exists it be only decent for it, for the sake of its scientific historian to put in an appearance with as little delay as possible.

A. O. Hume, *Stray Feathers*, 1873

In 1844 a new sparrow was described from Bahawalpur on the River Sutlej in the north of Sind. It took a further seven years after Hume's plea quoted above for the bird to be rediscovered by Scrope Doig in the Eastern Nara district of Sind. This bird was subsequently called the Sind Jungle Sparrow and has been found to occur in the valleys of the Indus and its four great tributaries, the Jhelum, Chenab, Ravi and Sutlej. It is restricted to a riverine habitat and was apparently overlooked for 36 years because of its close resemblance to a small House Sparrow.

Even today this is still a bird about which comparatively little is known. Odd reports suggested that its range might extend up the Sutlej into the Indian Punjab. Being in northwest India after our successful venture with the Cinnamon Sparrows gave us an excellent opportunity of putting this to the test and metaphorically killing two birds with one stone. We drove first to Ropar on the upper Sutlej, but a search along the river near there produced no sightings. The following day we proceeded downstream to Ludhiana where we were met by a guide laid on by the ever helpful Sugato. I explained that I wanted to get down to the bank of the river somewhere away from one of the few crossings. This was not as easy as it sounds and without our guide we should have certainly lost our way in the maze of unmarked dirt tracks threading the intensively cultivated farmland between the main road and the river, enquiring the way forward at each village we encountered. Everyone was most helpful and eventually we found the river near the village of Naniawal about 25 kilometres west of Ludhiana. The place looked ideal with scattered trees along the high containing banks and it was not long before we found our first Sind Jungle Sparrow, chirping in typical sparrow fashion from a kikur tree (an acacia). A number of nests were also found, but though the birds were calling beside them there was no sign that breeding was in progress.

That evening I visited the Punjab Agricultural University at Ludhiana and, though as always everyone was most helpful, nobody appeared to know anything about Sind Jungle Sparrows and a search through the library produced no more information about the bird than I already knew from my reading at home. The next day Sugato had arranged that we should go to the Harike Nature Reserve at the confluence of the Beas and the Sutlej, where it had been organized that the wildlife warden, Parmatma Singh Sandhu,

Party at Harike Reserve, Punjab, India: Parmatma Singh Sandhu, wildlife warden, Jaswani Rai, wildlife guard, and Manmit Singh, my guide.

would show us round. The warden had never heard of Sind Jungle Sparrows, but on getting a description of the habitat took me to a likely place and sure enough we quickly located a number of nests in a small sesame plantation and others in kikur trees standing in water. The birds were nest-building, but again there were no other signs of breeding activity.

By now it was late afternoon and the other less sparrow-oriented members of the party were turning their thoughts towards food. We bought a large fish from a stall beside the river and repaired to a rest house where we handed over the fish to the resident cook and, after dozing comfortably on gently fanned charpoys for about an hour, we sat down to an excellent meal to celebrate our success. I am not sure if I convinced the warden how lucky he was to have this comparatively rare sparrow on his reserve – his interest lay more in wildfowl and the larger mammals whose numbers have been greatly reduced by the increasing pressure of the human population since Partition – but at least he was pleased that I had found it. Subsequently, we searched other parts of the river and found the sparrow to be reasonably common in the immediate vicinity of the water as far upstream as Ladhowai, where the main road from Ludhiana to Jullundur crosses it. How far it extends from there towards Ropar, time did not allow us to investigate. Wherever there were small villages or habitations beside the river House Sparrows were also present, occupying the buildings, whereas the Sind Jungle Sparrows had nests in the nearby trees. One reason I had wanted to avoid the river crossings was that for security reasons the bridges have a permanent military guard and I thought that wandering around with field glasses and a large camera with a telephoto lens might cause trouble. On one occasion, when returning from a walk along the river to Ladhowai, I made a slight diversion in order to photograph a flock of Indian White-backed Vultures, leaving my wife to walk directly back to the bridge. When I caught up with her I found that she was already being entertained to tea by some of the off-duty guard staff. Conversation was difficult, but they were very friendly and did not seem concerned about my optical equipment. Birdwatching has always seemed to me to be an ideal cover for spying activities – perhaps that was why the Russians did not want me to go looking for sparrows in Turkestan near the rocket-launching station at Baikanour – but then I suppose a spy would not show himself as obviously as I did in sight of these guards.

Our guide in the Punjab, Manmit Singh, was a Sikh and most anxious that we should visit the Golden Temple at Amritsar, the Sikhs' holiest place of worship. The centre-piece of this magnificant holy complex is a floating golden shrine on a large artificial lake in the centre of the busy city. All visitors, having removed their shoes, are expected to take a drink of holy water at the entrance to the temple before going in. We explained that as visitors to India we had to be careful and not drink unboiled water. Manmit,

The Golden Temple, Amritsar.

however, assured us that being holy water it would be quite safe. We did not share his confidence, but to avoid offending him took the risk and had a drink; fortunately, there were no after-effects, though whether because its holiness extended to non-Sikhs or because much travelling through the country had by this time rendered us immune, I should not like to say.

Visiting India is an emotional experience for someone of British descent, tied up as it is with so much of our recent history. Signs of the British Raj are still plainly evident, though the *son et lumière* display at the Red Fort in Old Delhi does much to put the brief rule of the British Raj into its true historical perspective in the long and complex history of India. The dominant impression wherever one goes is of bustle: there are so many people, not only in the busy cities, but also in the surrounding country, with people everywhere seemingly on the move. The roads are so crowded with buses, lorries, cars, bullock carts, bicycles and the ever-present people that driving is a matter of continuous attention and skill. Mishra, our driver, appeared to know the width of the Ambassador – 95% of the cars on the roads seemed to be Ambassadors – to the nearest millimetre as he wound in and out of the other road users.

I was greatly impressed with the Punjab: with irrigation from the mighty Sutlej through vast systems of canals and the industry of the refugees from

Pakistan at the time of Partition, it has been turned from jungle into productive agricultural land. The only slight concern to me as a conservationist is that irrigation is now being brought to the neighbouring state of Haryana so that even more jungle will disappear under the blanket of agriculture. Punjab shares its state capital, Chandigarh, with Haryana, at least for the present, though this is something that is under discussion as the central government tries to solve the present problems with Sikh unrest in the Punjab. Chandigarh is a completely new city conceived by Le Corbusier to replace the former Punjab capital of Lahore, now in Pakistan. It is laid out on a generous scale – there is a lot of India – with an artificial lake, wide tree-lined streets, an enormous rose garden and, remarkably, an absence of cows, which are not allowed in the city; everywhere else they present another traffic hazard as they stroll about the main roads by night and day. We stayed one night there with yet another of Sugato Chaudhuri's contacts, Prem Parkash, and his most hospitable wife. He introduced us to Indian wine made from locally grown grapes, something I was quite unaware existed in India. Unfortunately, I gather that the venture has not been a commercial success so that it is an experience that is not likely to be repeated.

Let it not be thought that I am completely indifferent to culture, being in India for the first time we could not come home without visiting the Taj Mahal. I find that most of the famous sights that one has read about in

Mishra our trusty driver.

He stopped the car to rush out and pick up two Grey Francolin chicks when I would much have preferred to photograph them undisturbed.

advance tend to disappoint; the magical reality of the Taj is, however, far beyond expectation. It was really my long-suffering wife who had persuaded me that we should go to Agra to see the Taj. I had, of course, an ulterior motive. A quick glance at the map showed that on the way back to Delhi it would be convenient to call in at Bharatpur, the most famous of the Indian bird reserves. Its international reputation depends on it being a wintering place for one of the world's rarest birds, the Siberian Crane. This was high summer and the cranes and the other waterfowl for which the reserve is justly famous had long departed for their northern breeding quarters. Nobody considers Bharatpur worth a visit in the summer – we were the only visitors present at Forest Lodge – but nevertheless I found plenty of birds to interest me there during our all-too-brief 24-hour stop-over.

Later Sugato asked me how we had got on with our driver Mishra. Apart from minor language difficulties and his lack of understanding of our birdwatching interest – on one occasion he stopped the car to rush out and pick up two Grey Francolin chicks to present to my wife, when I would have much preferred to photograph them undisturbed from the car – we had got on very well with him. Sugato then told us that he had been suspected of murder over a land dispute on one of his leaves to his home village. Not that that had worried Sugato: disputes over land are certainly a serious matter in India, but he would not have expected Mishra to be anything but friendly with us.

Exotic dreams when sleeping on opium sacks.

14: Further east: Thailand

In Bangkok
At twelve o'clock
They foam at the mouth and run,
But mad dogs and Englishmen
Go out in the midday sun.
Noël Coward, '*Mad dogs and Englishmen*', 1932

In order to make sure that I made the most of my comparatively short annual leaves, the sparrow trips were well planned in advance, often by over a year before we actually set off. Another Asian species on my list was the Pegu Sparrow. This bird ranges widely over Vietnam, Cambodia, Laos, Thailand, Malaysia and Burma. In 1979, planning for a visit in 1980, it was not difficult to select Thailand as the most promising of these – the war was still going on in Vietnam and had spread to Cambodia and Laos, free movement for visitors to Burma is very restricted, and the expansion of range into Malaysia was fairly recent so that any observations made there would not necessarily be typical for the species. The pre-planning involves reading up the relevant literature and writing extensively for information. In this case I was greatly helped by Jeffery Boswall of the BBC who had recently made a series of nature films on Thailand and very kindly placed at my disposal a list of the ornithological contacts he had made in the country. The Thais are some of the nicest people I have met, but cannot be listed among the greatest correspondents. Dr Boonsong Lekagul, the doyen of Thai ornithologists, who is the co-author of an excellent book on the birds of Thailand, thought that Pegu Sparrows bred on his farm some 50 kilometres from Bangkok and very kindly invited me to stay there if I wanted to, but he was the only Thai who responded and, all in all, I was unable to get any specific information on the best place to go to be certain of finding the birds and also exactly when if I was to be sure of finding them breeding. I did get a number of leads from Americans who had worked in the country, mostly on overseas aid service, but these were a bit out-of-date as the individuals concerned were all back home in the States. The best lead came from Mrs Joy Adamson, a keen birdwatcher whose husband worked in the British embassy; she had seen some Pegu Sparrows recently in the nature reserve at Bang Phra, about 80 kilometres southeast of Bangkok. This seemed promising as Dr Joe T. Marshall, Jr, of the National Museum of Natural History in Washington, DC, had found Pegu Sparrows breeding at the Pasteur Serum Farm about 5 kilometres from the nature reserve, when he had been working on birds there in 1966 and 1967. I wrote to Dr Schwann Tunhikorn, the director of the nature reserve at Bang Phra for further information, but had no reply and having nothing really specific to go on decided that February–March seemed the most likely time to find the birds breeding. We planned to head for Bangkok and take it from there after we arrived. I wrote to Boonsong Lekagul saying we were going to arrive in Bangkok on 27 February and hoped we should be able to see him then. He replied by return of post saying he would be attending the regional meeting of the International Committee for Bird Preservation, Asian Continental Section, in Chiang Mai in the north of the country from 26 to 29 February and suggested that if we went straight there it would give us the opportunity of meeting most of the Thai ornithologists and finding out some more definite information on Pegu Sparrows. We immediately changed our flight to Chiang Mai, wrote off for hotel accommodation there and crossing our fingers departed without great confidence that it was going to be a successful expedition.

Things rapidly changed once we got to Chiang Mai, the 'Rose of the North', a pleasantly quiet city after the madhouse of Bangkok. Our letter had got to the hotel and a room was ready for us. What was even better the first person I met at that conference was Schwann Tunhikorn, who confirmed that Pegu Sparrows were present on the Bang Phra Reserve and invited us to stay there immediately after the meeting. He said he had not bothered writing as he was sure he would see me at the meeting in Chiang Mai. Oriental fatalism? My mind at rest, I proceeded to enjoy the conference (where I found I had become one of the official UK delegates!), the many interesting people we met there, not only from Thailand, but from much of southeast Asia as well, and the bird outings to the neighbouring hill country and the forest of Doi Pui in the Golden Triangle near the Burma border.

King Bhumipol was in residence at Phuping Palace at Buag Ha Mountain about 22 kilometres from Chiang Mai when we were there. This is not far from the troubled border region with Burma and there was plenty of evidence of soldiers on guard duty. At least this is what we presumed, when out early in the morning on birdwatching expeditions we saw them being dropped off along the road from army trucks in pairs. Evidently the duty extended only to the daylight hours as the trucks came out in the early evening and collected them again. Each pair had a sign like a small version of the 'lollipop' used by the attendants who shepherd schoolchildren across the roads in Britain. This was stuck in at the side of the road where they were dropped off, presumably to ensure that they were not overlooked at the collecting time if they had fallen asleep in the grass. We picked up one of the soldiers' caps that had dropped off on to the road; it was brand new with a large peak and has since served me very well as an excellent birdwatching hat. I trust the soldier was not severely punished for the loss.

Chiang Mai had been one of the areas I had intended to go to to look for Pegu Sparrows, but apart from some birds in cages in a wayside restaurant not far from Chiang Mai, we could find only one pair and a group of five nests at the edge of a field nearby. These nests were not in use. Oddly enough, though failing to locate many Pegu Sparrows, which everyone assured us were common in the area, at Doi Inthanon, we did see a flock of about 35 Chestnut Thrushes, a species that had been only recorded for the first time in Thailand earlier that year. Doi Inthanon was a fascinating place for birds, so on the spur of the moment a few of us decided to have an extra day up there spending the night out rough. Actually we stayed in a small hut and I had an excellent night sleeping on some sacks that I suspected had contained opium, this no doubt contributing to my deep sleep.

From Chiang Mai we returned to Bangkok by bus, a journey of about 700 kilometres that took nine hours. This gave us an opportunity of seeing something more of the country and was a most comfortable trip in an air-conditioned coach with cold drinks served by a hostess every hour or so, a stop for lunch at a restaurant, and a cold supper served on board, all for less than a modest £3. We were the only Europeans on board and in consequence were entertained by a variety of delicacies that were passed to us by our fellow passengers during the journey. The nine hours passed very quickly and the coach was infinitely preferable to the overnight train journey that would have

Schwann Tunhikorn, the director of the Bang Phra Reserve with the author's wife.

taken fourteen hours, most of it in the dark, or the return flight by air with nothing to see. My abiding impression, once we had descended through rather denuded forest south of Chiang Mai, was of flat paddy land and the *klongs* (as the canals are called) along the road filled with countless ducks. What happens to these I do not know as we were never offered duck on the menu.

Tree Sparrows were numerous in the busy streets of Bangkok and calling everywhere from the buildings in which they were nesting. Bangkok, of course, has much to offer apart from birds, not least the magnificent Buddhist temples. It is an incredibly busy city with numerous ten-lane highways that seemed to be filled at all times of the day and night with a mixture of lorries, cars, buses, rickshaws and bicycles changing lanes with careless abandon for some perceived advantage. Despite the disturbance, Bangkok offers a unique attraction: at night in the winter months large numbers of Swallows come in to roost on wires above the busy streets. Dr Samaisukh Sophason of Mahidol University in Bangkok, whom we met at Chiang Mai, had organized a count with some of his colleagues and came up with the astounding number of 200,000 birds.

From Bangkok, Schwann, who had returned from Chiang Mai by plane, drove us to Bang Phra, where we stayed in a delightful rest bungalow on the nature reserve. Not only was this very comfortable, but laid on was a pair of

Pegu Sparrows starting to build a nest in the roof. We breakfasted with the reserve staff and one morning my wife remarked on a large centipede that she had encountered during the night in our bathroom. Schwann enquired whether it had had a round or a flat body. My wife replied 'Flat'. With the unerring facility she has for these things she had come across one of a dangerous, poisonous kind. Schwann, who had lived on the reserve for five years, had never seen one there, but after breakfast was able to locate two in our bathroom after a thorough search.

The meals were unfamiliar by Western standards, consisting of dishes based on prawns, chicken, eggs, rice and so on arriving at the table in succession and all invariably accompanied by a delicious soup made from a chicken, prawn or fish stock flavoured with *makroot* (lemon grass) that was kept hot at the table on small oil lamps. To conclude there was always beautifully fresh tropical fruit. Meals were taken in an open-sided hut and we were usually joined by a Yellow-bellied Sunbird which was nesting outside and came in to watch us eating.

This was the pineapple season and I accompanied Schwann to a pineapple market. The fruit was in heaps in a large shed; on entering one was presented with a small rod like a conductor's baton with a knob on the end. This was for striking the fruit, ripeness being judged by the sound they emitted. I tried several heaps, but could detect no difference; either Schwann had a better ear or all the fruit was in perfect condition, because the ones we bought for a few pence each were always juicy and mouth-watering.

While staying at Bang Phra we were able to visit the serum farm and laboratory close by, where Joe Marshall had seen the Pegu Sparrows in the 1960s. The director, Dr Skon, not only placed the grounds at our disposal, but even had a fence removed so that we could drive the car on to one of the lawns where we could use it as a hide at a Pegu Sparrow's nest. The horse serum farm is a most interesting place. There are four poisonous snakes in Thailand. Venom is extracted from them at a snake farm in Bangkok that is a well-known tourist attraction and sent down to Bang Phra where it is injected into horses for the production of the serum. The horses there are ones that have retired from other duties and are well looked after and well fed in the very pleasant, shady surroundings. The venom has apparently little effect on the horses so that in retirement they live a most useful and at the same time pleasant existence that otherwise would have been unlikely to have been their fate. From Bang Phra the anti-venom is dispatched to clinics all over the country where, from the known distribution of the snakes and the different appearance of the puncture marks left by a bite, the appropriate injection can be given. Provided anyone bitten by a snake goes quickly to the nearest clinic, they should not be at hazard; the problem is rather of getting the information on the service across to a widely dispersed peasant population, but the effort to combat the problem of snake bite is a bold and imaginative one.

One was presented with a small rod with a knob on the end for striking the fruit, ripeness being judged by the sound they emitted.

As far as I was concerned the visit to Bang Phra was a rich and rewarding one. There were about 60 Pegu Sparrows on the reserve, some building nests on the reserve houses, others occupying the crowns of coconut palms. This was at the beginning of March and the birds were just at the start of their breeding cycle. At the serum farm, five kilometres away, the breeding season was further advanced with six pairs feeding young in nests in large mango trees. The earlier start of breeding in the latter place may have been because irrigation to provide good pasture for the horses had given an earlier flush of insects and other invertebrates that the sparrows depend on to feed their young.

Tree Sparrows were also present at both places and they were also only at the nest-building stage at the nature reserve, whereas they were feeding young at the serum farm. The two species of sparrows were living side by side, both using buildings on the reserve for nesting sites, while at the serum farm the Tree Sparrows were nesting on the office buildings and the stables with the Pegu Sparrows in the trees. The adults were visiting different areas, doubtless exploiting different sources of food for their young and apparently not in competition with each other for food or nesting sites.

Before returning home we decided to visit the Peninsula to have a look at a different part of Thailand. We selected Song Khla on the Gulf of Siam coast, mainly because of a convenient Thai Airways connection from Bangkok to nearby Hat Yai, but that night in the YMCA where we were staying, an American couple we met at dinner persuaded us that Phuket Island in the

An attractive Tree Sparrow nest on beach hut, Phuket Island, Thailand.

Andaman Sea was much more attractive. Arriving on Phuket Island late the following afternoon we spent the night in Phuket town and, on looking out of the bedroom window of our hotel the next morning, I saw a shop immediately opposite with a Hertz sign in the window. As soon as we had breakfasted we enquired about hiring a car, but ran into a difficulty – the sign had only been put up that morning and there were no cars ready. After a short discussion a car was promised and within a couple of hours we were on our way exploring the island. To my unexpected delight I located a colony of Pegu Sparrows at Hat Karon on the west coast of the island and, having found a simple Chinese-run beach hotel, decided to stay. This was a bonus as Pegu Sparrows had not previously been recorded from Phuket.

This appeared to be the only breeding colony on the island, though isolated birds were seen at other places, including one on the outskirts of Phuket Town. I counted 55 birds in the Hat Karon colony, an accurate count being possible as they roosted together at night in a large casuarina tree. These birds were at a later stage in breeding than the birds at Bang Phra, with fledged young out of the nests being fed by their parents, and they seemed about to start a second cycle of breeding judging by the way they were attending nests in sugar palms during the day. Tree Sparrows were inhabiting the same place, nesting in beach huts, but again there seemed to be no animosity between the two species.

One evening when we were having dinner in a beach restaurant we met a most interesting American army officer, Captain Daniel Macdonald, from the United States embassy in Bangkok. On being called up for military service he had gone to a school to learn Lao, following which he was posted to Thailand: beyond military matters all armies seem to have the same blinkered view of the outside world – Lao and Thai, although both Chinese languages in origin, are only distantly related. Captain Macdonald was not, however, discouraged and proceeded to learn Thai. He became so taken with the country that at the end of his period of military service he enlisted in the regular army and returned to Thailand as a member of the embassy staff. We spent a fascinating evening hearing about Thailand from a Westerner like ourselves, but one with a deep and passionate love for the country and its people.

Phuket is described as a holiday island and is much to be recommended to those looking for peace and quiet on hot, sandy beaches and a warm sea, unspoilt by the sophisticated accretions of so many developed holiday resorts. Most of the beaches have simple open-air restaurants where one can get cold drinks and freshly caught fish from the sea at incredibly low prices. All of these restaurants appear to be run by pretty girls in their teens and having served you the girls are happy to sit and talk with you – in English! – at the scrubbed board tables. At one such restaurant the girl who was serving apparently took a fancy to my wife and returned later with a small package made up from a page torn from an exercise book. After much unfolding a small gold rose on a fine gold chain dropped out. 'It is for you', she said and to my wife's 'Why for me?' she replied 'Because I like you'. Despite protests she hung it round my wife's neck and danced off to serve another customer. Remembering a piece of advice from our American acquaintance of the previous evening that non-acceptance of a gift from a Thai would cause grave offence there was little we could do about it and felt most ashamed as we had nothing to give in return, having taken the advice given to tourists by TAT (the Tourism Authority of Thailand) that nothing of value should be taken on to the beaches because of the risk of being robbed! This is the impression we should like to keep of Thailand, superficial though it may be after a stay of only a few weeks in the country – a beautiful place with friendly smiling people. Long may they keep from being embroiled in the sordid politics of that unhappy region of the world.

The Pegu Sparrow is a bird of open country with scattered trees and its penetration down the Malayan Peninsula must have been prevented by unbroken tracts of primary rain forest. However, this natural barrier has been removed by timber extraction providing corridors along which the Pegu Sparrow could spread; thus in 1938 they were recorded for the first time in northern Malaysia and since then have been moving gradually further south. Dr David Wells of the University of Malaysia, another contact from the conference at Chiang Mai, told me that the Pegu Sparrow is not particularly successful in peninsular Thailand and Malaysia so there may be other factors than purely the barrier of the rain forest that limit its range. Certainly it is difficult to explain why there was only one colony on Phuket Island which, on the surface, looked very suitable Pegu Sparrow country. No such inhibitions have acted in the case of the Tree Sparrow which is common both in Malaysia and Singapore. It is thought to have arrived there only in the 16th or 17th century via the coastal trading stations. Again this highlights the success of the two sparrows most associated with man, the House and Tree Sparrows, though I found it interesting that in the Pegu Sparrow there was yet another sparrow species that was quite happy to live in close association with man and nest on his buildings.

Want any help, Limey?

15: Failures:
USA, Seychelles, Venezuela, China

Not a whit, we defy augury; there's a special
providence in the fall of a sparrow

Shakespeare, *Hamlet,* V

Not all of the searches have had a successful outcome. The Tree Sparrow was introduced to St Louis, Missouri, in the United States, in 1889. I happened to arrive in St Louis on a business trip on a Sunday afternoon in 1959 and, having time to spare, I thought it would be nice to see if I could find some Tree Sparrows. The airport hotel was on the edge of the town and it was easy to walk out a short distance away along a highway to get into the surrounding country. I had bargained without the US police. I had got only a short distance before a police car drew up beside me and one of the cops lent out and asked 'What's the trouble, bud?' I had started to explain, when I was interrupted with the question 'You a Limey?' When I assented, he shook his head and the car pulled away. Only Limeys would go voluntarily for a walk in God's own country! I later found out that, although the Tree Sparrows had initially done well, when the House Sparrow moved in a few years later in the course of its expansion across the States the Tree Sparrow was displaced to the Illinois countryside north of the area I was in and I had been looking for them in entirely the wrong place. They are still in Illinois, though they have not done anything like as well as the House Sparrow, a recent estimate putting the total US population at only about 150,000 birds.

My next 'failure' was rather more exotic. In 1973 I heard that a flock of twenty House Sparrows had recently been seen in Victoria, the principal town on Mahé, in the Seychelles. Never having seen a recent colonization by House Sparrows, and moreover being rather attracted by the thought of going to the Seychelles, this seemed a good opportunity and off we went the following year. Despite searching Victoria, thoroughly, and that was not difficult – for a capital it is a pretty small place – we failed to find anything of the sparrows. Whether they had failed to colonize and had died out or the original sighting had been a misidentification was not clear. I had suspicions about the latter. Another introduced bird to the Seychelles that has been very successful is the Madagascar Fody. The male of this species is a brilliant red, but the hen is a typical little brown job that could easily be mistaken for a hen House Sparrow. But it was by no means a wasted trip; the Seychelles is a delightful place to visit without having to find any excuse and though lacking in sparrows has some other exciting birds.

We stayed in an old hotel, the Northolme, long since superseded by the new establishments that have grown up in recent years to cater for the tourist

A Seychelles beach. If you have to miss out with sparrows, where better to do it?

trade. The Northolme dated back to the days when the packet steamer from Bombay to Mombasa called in at the Seychelles and many expatriates returning home from India broke their journey there and stayed at the Northolme, including, we were told, Somerset Maugham who had written some of his stories there. We had a suite with a magnificent view over a turquoise sea to the island of Silhouette in the distance. Our rooms were built on stilts and the floorboards had long since shrunk so that there were gaps through which one could look down to the rocks below; awkward if one dropped jewellery through, but excellent for dispatching the enormous cockroaches and, more importantly, the centipedes – the latter being the only poisonous creatures on the island – both of which come out at night.

The following episode also involves the House Sparrow. It can only marginally be described as a 'failure'; I think I was perhaps a little premature to have hoped for success. I have already mentioned how the House Sparrow introduced to North America has spread south and has reached Panama. Similarly, birds introduced to South America have spread north with the spearheads now in Amazonia in the east and Colombia in the west. It seems inevitable that these groups will join up and Venezuela looks the most likely place. I was there in 1983 and again in 1988 and with my friend Tony Crease, now a resident in Venezuela, searched for House Sparrows near the grain silos in the port area of Maracaibo – a likely place for them to turn up – but to no avail. Neither could I find them in the capital Caracas, another possible spot.

It is a generally accepted hypothesis for speciation that groups of the same species, separated by some geophysical or climatic event, can diverge by adaptation to their different environments and differentiate to such an extent that, should they come together again, they are unable to interbreed freely and have in fact evolved to become separate species. Thus with the House Sparrow we have the unique possibility of this hypothesis being put to the test, though whether the 150 years or so the groups have been apart is long enough for sufficient evolutionary change to have occurred is rather doubtful. It is, however, such an unusual event that the opportunity must not be missed. I think this opportunity will certainly arise some time before the end of the century.

The final 'failure' is one that still concerns me. As I mentioned at the outset, there are twenty species of sparrows and my ambition was to see all of them in the field. I had succeeded with nineteen, but the twentieth, the Saxaul Sparrow, still eluded me. This bird lives in Russian and Chinese Turkestan. This was yet another reason why I wanted to go to Kazakhstan. Having failed in all my attempts to get into southern Soviet Asia and with relations with

China improving, I began to look at Chinese Turkestan as an alternative, at least as far as the Saxaul Sparrow was concerned.

Recent information on the distribution of the Saxaul Sparrow in China is somewhat sketchy, but as far as I could make out it is confined to an area in the northwest from Sinkiang through northern Kansu to Inner Mongolia. I identified a possible location. Western tourists are allowed to visit the Cave of a Thousand Buddhas near the desert town of Dunhuang in Kansu, between April and October. The way to get to Dunhuang is via Liuyuan, which lies on the rail line from Lanzhou to Ürümqi, and then take a bus from Liuyuan. Westerners are thus allowed to visit Liuyuan. According to one of the guidebooks for Western travellers 'There is a small dirty hotel in Liuyuan where one can stay overnight'. Liuyuan is in the desert right in the range of the Saxaul Sparrow. It seemed to be an excellent starting place with plenty of opportunity for nosing around once one had got away from the officialdom of the big tourist cities.

The next problem was to get to China. Wearing my engineering hat I managed to get invited to give a series of lectures that took in a visit to Xi'an. Xi'an, the ancient capital of China, is part-way on the rail line to Ürümqi – only two, or at the most three, days away by train. I suggested to my Chinese host, Xu Shaogau, General Secretary of the Chinese Mechanical Engineering Society, that I should like to end my lecture tour in Xi'an and from there take the train to Liuyuan. This caused a bit of consternation: the Chinese seemed to find it difficult to understand how an engineer could also have a professional interest in ornithology, but none the less he agreed to look into the possibility. A few days later the answer came back that it was not possible. Discreet enquiries revealed that, as a guest, my hosts would have had to organize the whole trip for me, including a guide to ensure that I did not get into any difficulties. Permission for this was requested from the Minister of Machine Building, a member of the Politburo, but unfortunately it was not forthcoming.

The only ornithological contact I was able to make in China was Professor Tso-Hsin Cheng at the Institute of Zoology, Academia Sinica, in Beijing. I had previously been in correspondence with Professor Cheng (about Saxaul Sparrows – what else?) and wrote that I hoped I should be able to see him when I was in Beijing. Even making this simple arrangement took three days, as there was no direct contact between the Chinese Mechanical Engineering Society and the Institute of Zoology and this had to be done through some circuitous bureaucratic route. I had a very interesting meeting with the professor and he was able to show me some Saxaul Sparrow skins that had been collected in Sinkiang, but no information on its present status. Professor Cheng, the doyen of Chinese ornithology, was then in his seventies, still very interested in birds and working a fourteen-hour day; although he had personally collected some of the Saxaul Sparrows that were in the Institute of Zoology, this had been in the 1930s. There have been considerable changes in China since then and quite possibly these have also affected the Saxaul Sparrows, particularly changes to the agriculture through improved irrigation that could have had an adverse effect on this semi-desert species.

One thing that interested me particularly about China was the status of the Tree Sparrow. From time to time the human population there has been mobilized into some sort of national activity: a campaign against spitting, reduction of litter, destruction of pests. Chairman Mao decreed that the sparrow was a pest and had to be exterminated. Han Suyin, the well-known writer, happened to be in Beijing at the time of the anti-sparrow war in 1958 and gave a graphic account in an article. Some days before the campaign was to start, the propaganda machine rolled into action in the newspapers and on the radio, while detailed instructions were issued over the street corner loudspeakers. It was put over that each sparrow consumed about 7 *catties* (4.5 kg) of grain per year and for every million sparrows killed sufficient food would become available for 60,000 people. Everyone had to co-operate in destroying the sparrows: they were to be poisoned, caught with bird lime or trapped and any nests had to be pulled out. The old and infirm had to join in by keeping the birds in the air by banging on tins and lighting fire crackers until they fell down exhausted – the sparrows, not the aged and infirm – and

The old and infirm had to join in by banging on tins and lighting fire crackers until the sparrows fell down exhausted.

could be dispatched by the younger and fitter Pioneers. The blaring of sirens at 04.45 on the appointed day marked the start of the action, which continued until dusk that day and for the following two days. According to the official statistics, 200,000 sparrows had been killed in Beijing by the end of the first day, 600,000 by the second, and a grand total of 800,000 birds at the end of the *battue*. This was repeated all over China, so the numbers killed must have been immense. Nowhere else in the world but China could have mounted an operation of this scale involving the whole population, and I was interested to see what had been the long-term effect.

There were plenty of sparrows in Beijing when I was there in 1984 and I was keen to learn more about what had happened in this ambitious control attempt. I asked Professor Cheng. The initial premiss had been a great over-simplification; Tree Sparrows feed insects to their young as well as consuming grain and thus are not entirely without benefit to man. The following year grain yields were down because of excessive insect pests and any further idea of killing off the sparrows was quickly dropped. The sparrow population returned to its normal level as the surviving birds bred and filled the gaps and the balance has been restored.

The Saxaul Sparrow remains my one real failure. Perhaps something will eventually turn up allowing me to visit one of their locations and I shall be able to complete my sparrow set. I can but hope.

The academics seemed unaware of what was going on around them.

16: More about the House Sparrow: Back home

Home is the sailor, home from the sea, and the hunter home from the hill.
R. L. Stevenson, '*Requiem*', 1887

The search goes on. One of the most remarkable features of the House Sparrow is its close association with man, not only can it be classed as a true commensal of man, dependent on him as a provider of food and nesting opportunities, but in places it has gone as far as to move in on man and live an under-cover existence. This is perhaps not so surprising in railway stations, stores and large factory buildings with easy access to the open air, where it may be joined by the feral Rock Dove or Town Pigeon, but it is also to be found in much more enclosed situations; for example, a small flock used to occupy the Queen's Building at London's Heathrow Airport, when this building served as the passenger terminal for European flights. These birds inhabited the area of the snack bar, resting in the concealed lighting recesses in the ceiling and coming down to feed on scraps both on the floors and on the tables themselves. As this was an air-conditioned building and as such cannot have provided easy access for the birds to the outside world, it is probable that they lived entirely indoors. I used to visit the Senior Common Room at Leeds University very regularly and was interested to see that there was usually a small party of House Sparrows in the dining room. These birds were equally as bold as the ones at Heathrow, feeding around one's feet and coming on to

R.N. S.P.D.C.(U.K.), EAGLESCLIFFE,
CO. DURHAM.

NOTICE

Sparrows in the Establishment

During recent months the number of sparrows which have taken up permanent residence in the various buildings has greatly increased, so much so that the birds are now considered to be a serious menace.

2. It is unfortunate that many of the buildings are constructed in such a manner that the birds can gain access comparatively easily, not only through doorways, which must necessarily remain open for long periods, but also through holes in the side walls in way of roof trusses etc. Having once become accustomed to a particular building the birds build nests and, if left undisturbed, they and their descendents are then there for good.

3. The situation has now been reached where the noise and smell and the general fouling of the inside of buildings are intolerable and it has therefore been decided to take action forthwith to endeavour to eradicate the birds. This decision was unanimously agreed upon at a meeting of the Yard Whitley Committee held on 11th August, 1953.

4. The method of tackling the problem will be as follows:-

(i) Works Department will be requisitioned to remove all nests from the insides of buildings. This it is hoped will unsettle the birds and render them less likely to wish to return to a particular building.

(ii) An endeavour will be made to trap as many birds as possible in a special cage. These birds will be taken away to a good distance from the Establishment and then released.

(iii) As many access holes in the sides and roofs of buildings as it is possible to seal will be blanked off.

5. If the above measures are successful, it will then be necessary to continue to do everything possible to discourage the birds from gaining another footing. It is hoped that everyone will co-operate in this undertaking, which is necessitated by considerations of normal hygiene, and that they will neither hinder the work nor encourage the birds to remain.

CAPTAIN (E)-IN-CHARGE

14th August, 1953.

Naval order of battle against the House Sparrow.

the tables. It always surprised me that my hosts never remarked on this and when I tried to get further information – and what better place to get information than within a society of academics – nobody seemed to know how long they had been there nor in fact anything about them at all.

A more amusing episode concerned a local naval stores depot. Sparrows became so numerous in the stores buildings that there were complaints from the trade unions that the workplace was unhygienic and the Clerk of Works, Percy Rayfield, who happened to be a birdwatching friend of mine, was instructed to do something about it. He constructed a large trap and, as he was unwilling to kill the birds, I suggested that he should transport them about ten miles away before release. Transplanting experiments in this country have shown that few, if any, House Sparrows have returned to their place of origin when taken as far as this. There was, however, an unexpected problem: the British are probably more soft-hearted about birds than any other nation in the world and in the works there was a subversive underground faction that did not approve of the sparrows being trapped. Mysteriously, the cage door was found to be open after it had been carefully shut and baited, and eventually the project was abandoned. Nature, however, came to the rescue; the following winter a Tawny Owl took up residence in the stores buildings and lived sumptuously on an easy prey. I was reminded of the advice given by the Duke of Wellington to Queen Victoria when she asked what could be done about the sparrows in the Crystal Palace. 'A sparrowhawk, Ma'am', was the reply.

Percy Rayfield was a good friend of mine and he not only supplied me with numerous acute observations but also provided me with the young House Sparrows that I reared and kept in an aviary in my garden in order to make closely controlled observations in connection with my studies on this species. These birds were only two days old when I got them and had to be fed every 10–15 minutes. I took them to work with me and kept them in a drawer in my office desk. When they got hungry they became very noisy demanding to be fed. One day my boss was in talking to me and after about 10 minutes the chirping from my desk became so intense that even he, an engineer quite uninterested in natural history, began to notice. I had to put him off by saying that they were nesting in one of the ventilators.

Percy's steady stream of penetrating comments on House Sparrows continued after his transfer to the Royal Naval Dockyard at Portsmouth, though I must confess that I was somewhat sceptical about the new race of 'naval' sparrows that he suggested was evolving in Royal Naval ports.

Of even more interest to me than sparrows in buildings were the tantalizing reports that I came across from time to time of House Sparrows down coal mines. The first of these was a brief note in 1956 in one of the Sunday papers reporting: 'House Sparrows were nesting hundreds of feet under-

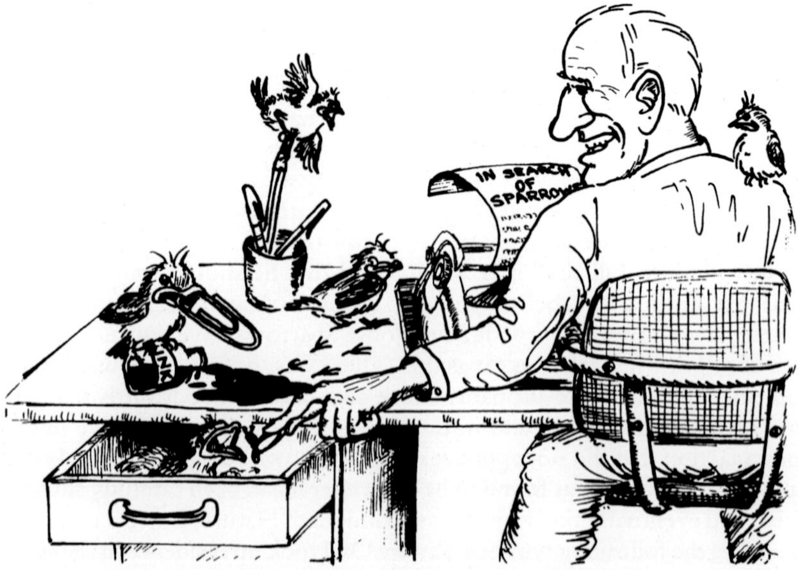

I took them to work with me and kept them in a drawer.

ground at the bottom of the pit shaft at Linton Colliery in Northumberland, where they were being fed by the miners.' By the time this item was drawn to my attention the trail had gone disappointingly cold and I was unable to get any more information.

Six years later, on 21 March 1962, I was driving home from work when I heard an item on the BBC Northern News that House Sparrows were building a nest at the floor of the shaft at Waldridge 'B' pit in County Durham, 180 metres below ground level. I wrote at once to the Area Manager of the Coal Board, but unfortunately it took some time for my letter to reach the appropriate Mine Manager, who by the time he received it had already given instructions that the birds should be caught and released above ground. The birds, a cock and a hen, were first seen on 12 March and they were fed on bread and crushed oats from the mine stables; water was available on the mine floor. Although there was plenty of hay and straw around, the Mine Manager said there was no evidence of the birds attempting to build a nest – I suppose the news item would not have been worthy of inclusion without this little bit of embellishment by the reporter. The hen was caught and brought to the surface on the night of 29/30 March, but the cock was not seen again. These birds were never more than 150 metres from the shaft bottom, an area that was continuously lit by electric light except for a 24-hour

Naval sparrow.

period from 6 a.m. on Saturdays. Waldridge 'B' was a mine with pit ponies and it seems likely that the sparrows had entered the pit cage for hay seeds and corn spilt on the floor and got unwittingly carried underground. The manager knew of two earlier occasions of sparrows down in the pit, but with no evidence of them having survived for any length of time.

The next report was from a more reliable source than the news media – a Canon of the Church. This was from Horden Colliery, also in County Durham. Only one bird was involved so there was no question of breeding, but the bird was apparently living happily at a depth of 300 metres with plenty of food supplied by the miners.

These reports were just appetizers to the one that appeared in the *Daily Telegraph* on 5 November 1977 and was more specific; it was that a pair of House Sparrows was actually feeding young in a nest nearly 650 metres below ground at Frickley Colliery in South Yorkshire. A letter to the Mine Manager confirmed that the birds were genuinely breeding and I was very kindly given permission by the Area Manager of the Coal Board to visit the scene of the action. This took some time to organize and it was not until 10 March of the following year that I was able to go down the pit. The mine staff could not have been more helpful. Suitably clothed I was provided with a guide and we descended with a group of miners. The small town of South Emsall in which Frickley Colliery is situated consists of two quite separate communities – the miners and the rest. The mining community is completely self-contained with a distinct dialect, almost a different language. The miners were obviously well aware of the reason for my visit and I was clearly the object of a good deal of chaff on the way down. Unfortunately, with their strong dialect I was unable to follow what they were saying and thus could not reply in kind.

The story of the sparrows was as follows. Two birds were first seen at the bottom of the mineral shaft in June 1975 by the onsetter, Joe Kemp. The onsetter is normally the only man in this part of the mine workings and Joe Kemp and his colleagues, Herbert Coates and Joe Howcroft, kept the birds supplied with food and water. The following June a third bird arrived and joined the first two. It was evident that at the time of their entrance to the pit they were all recently fledged, rather inexperienced birds that had settled on an empty skip when it was on the surface – I noticed during my visit that there were a lot of sparrows near where the skips came up. It appears that the first two birds were males and the third a female. In the autumn of 1977 the birds were seen picking up scraps of material and the observant Joe Kemp immediately took them down some grass and feathers, which the birds quickly used to build a nest on one of the roof girders. By the beginning of November they were seen to be feeding young in the nest and subsequently three young birds fledged, though unfortunately they did not survive. I think the most probable reason for this is that the adults were only provided with bread and seeds, whereas it is important for healthy development that young sparrows receive animal food for at least the first three days of life. Although the adults had been seen from time to time catching moths that had strayed down the pit, there were obviously not enough insects available for them to get sufficient for the young. When I visited the mine on 10th March 1978, I was able to see only one male, though I was assured that all three adult birds were still there; despite their long sojourn in confined conditions they were still exceedingly wary and obviously very healthy. The nest was still visible behind a roof truss.

The area of the mine in which the birds lived was a tunnel, 6 metres high by 9 metres wide with continuous illumination. During the week an onsetter is there 24 hours per day, but it is deserted at weekends unless there is any maintenance work to be carried out. It was amusing to find a bird table well stocked with food in this unlikely spot.

I found it of interest that the birds had started to breed in October, whereas the usual start for a young pair of House Sparrows in South Yorkshire would be April or May. Birds in temperate latitudes time the seasons by change in day length and the beginning of the breeding season is triggered by increasing temperature and the availability of suitable food for the young. No such clues were available, however, for the sparrows in the mine, where there was continuous illumination and a steady temperature of 15–20°C. It has been found that human beings kept underground for prolonged periods without any visual clues to what is happening above, tend to underestimate the passage of time and when they resurface are surprised to find how many days have elapsed. The fact that the sparrows delayed their breeding until October points to a similar phenomenon with the birds, a simple calculation suggesting that they were operating on a 35-hour day.

I suggested to Joe Kemp that if the sparrows attempted to breed again he should put out maggots at the appropriate time to give the young a better chance of survival. This was not to be, as by September 1978 there was only one remaining. However, it is remarkable that they should have got as far as they did and that at least one lived for over three years underground, a good lifespan for a House Sparrow living in normal conditions.

The final word of the Frickley sparrows must come from the birds themselves, as reported by Mr D. Southall of Doncaster:

The Miner Birds

Two sparrows went exploring, far from town
Curiosity proved fatal and they were sucked right down
To a world of coal, and Miners, beneath fluorescent lights
In other words at Frickley, they'd come to see the sights.

We are so disillusioned, in this world one sees
No other birds no sun no trees
But one bright feature, there is we find
Some mucky faced Miners who are so kind.

They feed us with seed, and lumps of fat
Ala pit bottom, and we're fond of that
News gets around, that we're being fed
So another sucker joins us half bloody dead.

They try to catch us, but we just fly
Wonder if they want us, for a tasty sparrow pie
Or could it be, they want us to set us free
In that world of sun, green fields and a tree.

So thanks to Messrs. Hepworth, J. Kemp, Herbert Coates
For a simple kindness that ever floats
In this world of discord so rare indeed
May the enclosed provide a stock of seed.

Joe Kemp, in addition to his duties as onsetter, was also responsible for the Canaries at Frickley, the Canary centre for the Yorkshire Coal Board Area. He was devoted to his birds and proudly showed them to me with complete details of the life history of each individual. It is interesting that even in those days the Canary was still maintaining its place as a sensitive detector of gas in mines, though since then high technology has finally taken over and the Canaries have been pensioned off. The miners had a great affection for them and I was pleased to learn that, even if the birds did succumb to gas, they recovered quickly if removed to a pure atmosphere and apparently suffered no permanent ill-effects. The sparrows that went down Frickley were extremely lucky to have chosen the pit where the bird lover Joe Kemp was at hand to look after them.

Thus ends my final search. Less than 150 kilometres from home, but 650 metres below the ground!

The miner birds.

Postlude

If he is content to take his information from others he may get through this book with little trouble.

Samuel Johnson

This is not the appropriate place to discuss the conclusions I have reached on the evolution of the sparrows. That has been done elsewhere, though those interested will have spotted a number of clues in the text. I do hope, however, that I have managed to convey something of the pleasure, even excitement, to be obtained from watching birds with an objective, rather than merely ticking off the species identified on a list, though I would not scorn this and the harmless pleasure it gives to many – I do it myself.

I should, however, like to draw attention to one thing that has forcibly struck me as a result of these studies. In our brief personal snapshot of 50 years or so of the natural scene, we tend to overlook the dynamism of animal species. The changes in distribution that have taken place in the species of one genus of birds serve to remind us of this dynamism. Within recent memory the increase in range of the House Sparrow, both natural and man-assisted, has been remarkable. And the less well-known changes in Pegu, Dead Sea and Desert Sparrows all reinforce the point that the living scene is under constant change. The problem in many cases is to account for the changes that are taking place, but at least birdwatchers the world over can make their contribution by providing information on the changes they observe. Do not neglect to observe the humble sparrows!

2 September 1991

Appendix A

SCIENTIFIC NAMES OF BIRDS MENTIONED IN THE TEXT

Bittern	*Botaurus stellaris*
Booby	*Sula* sp.
Bustard, Houbara	*Chlamydotus undulata*
Buzzard	*Buteo buteo*
Canary	*Serinus canaria*
Chukar	*Alectoris chukar*
Crane, Siberian	*Grus leucogeranus*
Dove, Rock	*Columba livia*
Finch, Darwin's	*Geospizidae*
Snow	*Montifringilla nivalis*
Flamingo	*Phoenicopterus* sp.
Fody, Madagascar	*Foudia madagascariensis*
Francolin, Grey	*Francolinus pondicerianus*
Goldfinch	*Carduelis carduelis*
Goshawk	*Accipiter gentilis*
Grackle, Tristram's	*Onycognathus tristramii*
Kingfisher, Black-capped Purple	*Halcyon pileata*
Eurasian	*Alcedo atthis*
Grey-headed	*Halcyon leucocephala*
Stork-billed	*Pelargopsis capensis*
White-breasted	*Halcyon smyrnensis*
Magpie, Ceylon Blue	*Urocissa ornata*
Pigeon, Town	see Dove, Rock
Raven, Brown-necked	*Corvus ruficollis*
Shrike, Great Grey	*Lanius excubitor*
Sparrow, Cape	*Passer melanurus*
Cape Verde Rufous	see Sparrow, Iago
Cinnamon	*Passer rutilans*
Dead Sea	*Passer moabiticus*
Desert	*Passer simplex*
Golden	*Passer luteus*
Great	see Sparrow, Rufous
House	*Passer domesticus*
Iago	*Passer iagoensis*
Italian	*Passer hispaniolensis italiae*
Pegu	*Passer flaveolus*
Rock	*Petronia petronia*

Sparrow—*continued*	
Rufous	*Passer motitensis*
Saxaul	*Passer ammodendri*
Sind Jungle	*Passer pyrrhonotus*
Somali	*Passer castanopterus*
Spanish	see Sparrow, Willow
Swainson's	*Passer swainsonii*
Tree	*Passer montanus*
Willow	*Passer hispaniolensis*
Sparrowhawk	*Accipiter nisus*
Starling	*Sturnus vulgaris*
Stork, Painted	*Ibis leucocephalus*
Sunbird, Yellow-bellied	*Nectarinia jugularis*
Swallow	*Hirundo rustica*
Thrush, Chestnut	*Turdus rubrocanus*
Tit, Black-crested	*Parus rubidiventris*
Vulture, Indian White-backed	*Gyps bengalensis*
Wallcreeper	*Tichodroma murina*
Warbler, Tristram's	*Sylvia deserticola*

Appendix B

BIBLIOGRAPHY

Bannerman, D. A. & Bannerman, W. M. 1968. *History of the Birds of the Cape Verde Islands*. Edinburgh & London: Oliver & Boyd.

Carruthers, D. 1949. *Beyond the Caspian*. Ebinburgh: Oliver & Boyd.

Cody, M. L. 1974. *Competition and the Structure of Bird Communities*. Princeton, NJ: Princeton University Press.

Dissanayake, Chandra. 1976. *Ceylon Cookery*, 2nd Edn. Colombo: Private Publication.

Fleming, Robert, L., Sr., Fleming, Robert, L., Jr. & Bangdel, Lain Singh. 1979. *Birds of Nepal*, 2nd Edn. Kathmandu: Avlock.

Flint, P. R. & Stewart, P. F. 1983. *The Birds of Cyprus*. London: British Ornithologists' Union.

Jenninga, Michael, C. 1981. *The Birds of Saudi Arabia: a Check-list*. Cambridge: Private Publication.

Lekagul, Boonsong & Cronin, Edward, W. 1974. *Bird Guide of Thailand*, 2nd Edn. Bangkok: Kurushpa.

Newton, Alfred. 1896. *A Dictionary of Birds*. London: Adam & Charles Black.

Phillips, W. A. 1975. *Annotated Check List of the Birds of Ceylon*. Colombo: Wildlife and Nature Protection Society of Ceylon.

Summers-Smith, J. D. 1963. *The House Sparrow*. London: Collins.

Summers-Smith, J. D. 1988. *The Sparrows*. Calton, Staffs: Poyser.

Suyin, Han. 1959. 'The sparrow shall fall', *The New Yorker*, 10 October.

Index